土木工程科技创新与发展研究前沿丛书

Mining Environment and Rockmass Control in Deep Coastal Mine
滨海矿山深部开采环境与岩层控制

Guo Qifeng, Ren Fenhua, Ma Minghui, Xi Xun

郭奇峰　任奋华　马明辉　席　迅　编著

中国建筑工业出版社

图书在版编目（CIP）数据

滨海矿山深部开采环境与岩层控制 = Mining Environment and Rockmass Control in Deep Coastal Mine：英文 / 郭奇峰等编著. — 北京：中国建筑工业出版社，2021.4
（土木工程科技创新与发展研究前沿丛书）
ISBN 978-7-112-26056-0

Ⅰ. ①滨… Ⅱ. ①郭… Ⅲ. ①沿海-矿山开采-矿山环境-研究-山东-英文②沿海-矿山开采-岩层控制-研究-山东-英文 Ⅳ. ①X322.252②TD325

中国版本图书馆 CIP 数据核字（2021）第 066915 号

责任编辑：聂 伟 吉万旺
责任校对：张惠雯

土木工程科技创新与发展研究前沿丛书
Mining Environment and Rockmass Control in Deep Coastal Mine
滨海矿山深部开采环境与岩层控制
Guo Qifeng，Ren Fenhua，Ma Minghui，Xi Xun
郭奇峰 任奋华 马明辉 席 迅 编著

*

中国建筑工业出版社出版、发行（北京海淀三里河路 9 号）
各地新华书店、建筑书店经销
北京鸿文瀚海文化传媒有限公司制版
北京中科印刷有限公司印刷

*

开本：787 毫米×960 毫米 1/16 印张：15½ 字数：310 千字
2021 年 12 月第一版 2021 年 12 月第一次印刷
定价：**68.00** 元
ISBN 978-7-112-26056-0
(37647)

版权所有 翻印必究
如有印装质量问题，可寄本社图书出版中心退换
（邮政编码 100037）

Brief Introduction

This book presents theories, methods, and technologies to ensure the efficiency and safety of deep coastal mining in Sanshandao Gold Mine, the largest coastal metal mine in China. The book is divided into 7 chapters, including opportunities and challenges in deep mining, geological conditions in Sanshandao mining area, in-situ stress measurement method and its application, rock mass investigation and rock quality evaluation, rock chemical damage induced by deep hydrochemical environment, mining optimization and stability analysis of stopes, and stability analysis and ground pressure monitoring.

This book can be used as a reference for senior undergraduate and graduate students in rock mechanics, mining engineering, civil engineering, hydraulic engineering, geotechnical/foundation, and geoscience. This book is also a good reference book for students from oversea, especially those from the countries of the Belt and Road. It can also be used as a technical reference for mining and geotechnical engineers working on the design, optimization and safety of deep mining.

本书以我国最大的滨海金属矿山三山岛金矿为背景,介绍了保障深部安全高效开采相关的理论、方法和技术。全书分为7章,包括深部开采的机遇与挑战、三山岛矿区工程地质、地应力测量及其实践、岩体测试与岩体质量评价、深部水化学环境水诱发岩石化学损伤、深部采场优化和稳定性分析、稳定性分析及地压监测。

本书可作为高等学校岩石力学、采矿工程、土木工程、水利工程、岩土工程和地球科学专业的高年级本科生和研究生,尤其是来自"一带一路"沿线国家留学生的教学参考书,也可供从事深部采矿的设计、优化和安全的采矿和岩土工程师参考。

■ Preface ■

The increasing demand for minerals, including metals and coals, and the exhaustion of shallow mineral resources make mining activities deeper and deeper. It has been reported that there are 112 underground metal mines with a mining depth of over 1000 m in the world. Ranking by the number of deep mines, the top five countries are Canada, South Africa, Australia, the United States, and Russia. Among these 112 deep mines, 58 are with a mining depth of 1000-1500 m, 25 with a mining depth of 1500-2000 m, 13 with a mining depth of 2000-2500 m, and 16 with a mining depth of 3000 m. More than 70% of them are gold and copper mines. For more than one-third of non-ferrous metal mines in China, the mining depth will exceed 1000 m in the next 10 years. It can be expected that deep mining will become a common issue for underground mines in the near future.

However, deep mining occurs in a critical environment with high in-situ stress, high geotemperature, high water pressure, and strong excavation-induced disturbance, posing significant challenges to the efficiency and safety of deep mining. Therefore, there is an urgent need to measure the deep mining environment, understand deep rock behavior, and optimize mining methods for deep mines. The objective of this book is to present theories, methods and technologies for improving the efficiency and safety of deep coastal mining, and particularly introduce the mining environment measurements, the coupled chemical-mechanical behavior of hard rock, and the optimization of mining methods in deep mines.

Sanshandao Gold Mine is not only a typical deep mine with a maximum excavation depth of 1200 m but also a large coastal mine with the mineral deposits and faults extending to the seabed of Bohai Sea. With the increase of mining depth, the mining conditions are becoming harsher. The risks of stopes collapse, roof failure and rockburst caused by high in-situ stress and high water pressure increase, which significantly affects the mining efficiency and safety in Sanshandao Gold Mine. Therefore, new and comprehensive techniques and methods must be employed to ensure the safety in the deep mining area.

This book firstly introduces the current status and future development of deep mining. Taking Sanshandao Gold Mine as a worked example, the geological conditions, especially for the high in-situ stress field, are first investigated and measured. Then the physical and mechanical parameters for rock mass in different mining depths are tested and analyzed. The mining activity in Sanshandao Gold Mine is significantly affected by the embedded seawater. Therefore, the chemical damage mechanisms of rock attacked by seawater are studied by experiments and theories. Further, the mining methods for deep deposit exploration are compared and optimized through numerical simulations. Finally, the stability of tunnels in the fault damage zone is modeled and monitored.

The authors would like to express their deep gratitude to Professor Meifeng Cai, academician of the Chinese Academy of Engineering, for his support on this book's publication. The authors are also very thankful for contributions from Dr. Yuan Li, Dr. Peitao Wang, Dr. Dong Ji, Dr. Xu Wu, Dr. Ying Zhang and Dr. Jiliang Pan, etc. Due to the authors' knowledge, some mistakes or gaps may unavoidably exist in the current version. We welcome comments from readers.

序 言

随着浅部资源的逐渐枯竭和不断增加的对矿产资源的需求，矿井开采深度正在不断增加。据统计，全世界开采深度 1000 m 以上的地下金属矿山共 112 座。按深井数量排名，处于前 5 位的国家是：加拿大、南非、澳大利亚、美国、俄罗斯。在这 112 座深井矿山中，开采深度 1000～1500 m 的 58 座，1500～2000 m 的 25 座，2000～2500 m 的 13 座，超过 3000 m 的 16 座。其中，70% 以上为金矿和铜矿。在未来 10 年左右，我国将有三分之一的有色金属矿山开采深度达到或超过 1000 m。可以预计，深部开采即将成为资源开发所面临的普遍问题。

深部开采面临高应力、高地温、高水压和强开采扰动的"三高一扰动"严苛环境，对安全高效采矿带来了严峻挑战。因此，测试深部开采环境、揭示深部岩石行为、优化深部开采方法对金属矿深部安全高效开采十分重要。本书旨在介绍保障深部安全高效开采相关的理论、方法与技术，特别介绍深部开采环境测试、硬岩化学损伤、采矿方法优化等方面的研究成果。

三山岛金矿采深已超过 1200 m，其不仅是我国典型的深部金属矿，同时又是我国最大的滨海金属矿。随着开采深度的增加，开采条件持续恶化。高地应力、高水压引起的采场塌方、冒顶、岩爆等事故隐患日益增多，严重影响三山岛金矿深部安全高效开采。因此，在三山岛金矿深部开采中必须采取新的综合技术以确保矿山安全。本书首先介绍了深部开采中的机遇与挑战，随后以三山岛金矿为背景，分析了三山岛矿区工程地质条件，进行了深部地应力测量研究及实践，开展了深部岩石力学特性与质量评价，考虑地下古海水的存在，通过理论和试验研究了岩石化学损伤破坏机理，然后对深部采场设计进行了优化分析，最后分析了裂隙岩体的稳定性和地压监测。

编者们深深感谢蔡美峰院士在此书出版中给予的帮助，同时感谢长期以来共同开展科研工作，为此书出版做出贡献的同事同学：李远、王培涛、冀东、武旭、张英、潘继良等。编者水平有限，书中错误和疏漏之处在所难免，敬请广大读者批评指正！

Contents

1 **Opportunities and Challenges in Deep Mining** 1
　1.1　Status of deep mining 1
　1.2　Mechanical environment in deep mining 2
　1.3　Engineering response in deep mining 3
　1.4　Status of deep metal mines at home and abroad 4
　1.5　Challenges in deep mining 8

2 **Geological Conditions in Sanshandao Mining Area** 15
　2.1　Regional geology 16
　2.2　Engineering geology 18
　2.3　Hydrogeology 23

3 **In-situ Stress Measurement Method and Its Application** 30
　3.1　The principle of in-situ digitizing hollow inclusion cell 30
　3.2　In-situ stress measurement and distribution law 35
　3.3　The in-situ stress field distribution 40

4 **Rock Mass Investigation and Rock Quality Evaluation** 49
　4.1　Influence of structural plane on surrounding rocks 49
　4.2　Structural plane investigation of deep tunnel rock mass 50
　4.3　Investigation summary of the structural plane of deep tunnel rock mass 55
　4.4　Rock quality classification 63

5 **Rock Chemical Damage Induced by Deep Hydrochemical Environment** 84
　5.1　Analysis of deep hydrochemical environment 84
　5.2　Damage mechanism and time-dependent characteristics of granite under the action of acidic chemical solutions 102
　5.3　Morphology and porosity development of granite before and after chemical erosion 121
　5.4　Time-dependent characteristic test of granite damage process under the action of acidic chemical solution 131
　5.5　Mineral composition change and chemical damage mechanism during rock-groundwater interaction 141

6	Mining Optimization and Stability Analysis of Stopes	149
	6.1 Optimization of deep mining method	149
	6.2 Optimization of structure parameters in deep stope	156
	6.3 Spatio-temporal variation of displacement and stress in stope	168
	6.4 Spatio-temporal variation of displacement and stress in multiple mining section mining	175
	6.5 Stability analysis of surrounding rocks in deep fault zone	181
7	Stability Analysis and Ground Pressure Monitoring	189
	7.1 Three-dimensional fracture network model	189
	7.2 Mechanical parameters of representative elementary volume	190
	7.3 Stability analysis of the tunnel in fracture rock mass	200
	7.4 Monitoring of ground pressure in deep mining	214
References		227

1

Opportunities and Challenges in Deep Mining

1.1 Status of deep mining

With the depletion of shallow resources, the deep resources mining has begun in China and other countries in succession. Since the 21st century, the Ministry of Science and Technology, the National Natural Science Foundation of China, the Ministry of Education and other institutions have approved many research projects on the basic theories and applied technologies of deep deposit mining. The State Council has approved the project of "Crisis Mine" with an expenditure of 4 billion Chinese Yuan, and proposed the principle of "focusing on prospecting in the periphery and deep of the mine", showing the importance to the issue of deep resources. In 2009, the Chinese Academy of Sciences has published the Roadmap for China's 2050 Science and Technology Development, which proposed the "China Four Thousand Meters Underground Transparency Program".

The deep deposit mining involves a series of issues, such as the concept of deep, key scientific theories and technologies, development trend and prospect. There is no unified standard for the mining depth of deep mining at home and abroad till now.

The mining industry of the western United States, marked by the discovery of the Black Mountain Gold Mine in 1874, has divided its mining into two periods: the shallow mining period and the deep excavating period. In terms of deep, it is usually defined as mining at a depth of 5000 feet (1554 m) or more. South Africa refers to mines of 1500 m as deep mines. Russian scholars have divided the deep mines into two types: one is the dichotomy, the mine shaft with the depth of 600-1000 m is called deep mine, the mine shaft with the depth of 1000-1500 m is called extremely deep mine; the other is the trichotomy, the mining depth of more than 600 m is referred to as deep mine, in which the first type of mine shaft is with the depth of 600-800 m; the second type of mine shaft is with the depth of 800-1000 m; the third type of mine shaft is with the depth over 1000 m. Japan has a critical depth of 600 m, while the UK and Poland have a critical depth of 750 m.

There are various classification schemes for the partitions of deep mines in China. In the coal system, some scholars divide the shaft depth of vertical mine into five categories according to the difficulty of sinking technology and equipment. Shallow mine is less than 300 m, medium-deep mine is 300-800 m, deep mine is 800-1200 m, ultra-deep mine is 1200-1600 m, and super-deep mine is more than 1600 m. In the metallurgical system, researchers divide the mining depth into four categories: shallow mining is less than 300 m, medium-deep mining is 300-600 m, deep mining is 600-2000 m, and ultra-deep mining is more than 2000 m. China's Mining Manual stipulates that the mining depth of 600-900 m is deep mining, and the depth over 2000 m is ultra-deep mining.

Xie Heping, an academician of CAE, believes that the concept of deep shall comprehensively reflect the stress level, stress state and surrounding rock properties of the deep. Deep mining is not based on a depth, but a mechanical state. The three concepts of subcritical depth, critical depth and ultra-deep critical depth of deep mining are put forward by considering three factors of stress state, stress level and properties of coal and rock mass integrally, and the specific calculation formulas are also given.

Although the definition depth of deep mine shaft proposed by various countries is not unified, there is a general threshold value in each country. For example: 600 m for Japan and Russia, 700-750 m for Britain and Poland, 900 m for Germany, 1000 m for Russia, 1500 m for South Africa and 1550 m for the United States. According to the current status and future development trend, combined with the objective reality of mine mining in China, most experts believe that the initial depth of deep mining in China can be defined as: 800-1000 m for coal mine and 1000 m for metal mine.

1.2 Mechanical environment in deep mining

In deep mining, the mechanical environment of rock mass changes, which leads to different engineering responses. Comparing with shallow mining, deep mining has the characteristics of "three high", i. e. high in-situ stress, high geotemperature and high water pressure.

(1) High in-situ stress

The rock mass in deep environment bears the vertical stress caused by the dead weight of the overlying rock stratum and the tectonic stress caused by the geological conformation movement. When the occurred depth reaches thousands of meters, there will be a huge primary rock stress field in the rock mass. According to the in-situ stress measurement of South Africa, the in-situ stress level is 95-135 MPa at the depth of 3500-5000 m. Thus, the deep rock mass has an abnormally high in-situ stress field and accumulates huge deformation energy.

(2) High geotemperature

According to measurements, the deeper the undergroundis, the higher the geotemperature is. Geotemperature gradient is generally 3-5℃/100 m. In some areas, such as near faults or abnormal local areas with high thermal conductivity, the geotemperature gradient may be as high as 20℃/100 m. The temperature change of 1℃ in rock mass might produce the in-situ stress change of 0.4-0.5 MPa. Therefore, the high geotemperature of deep rock mass will exert a significant influence on the mechanical properties of rock mass, especially the rheological and plastic instability of deep rock mass under the condition of high stress and high geotemperature is greatly different from that under the condition of ordinary environment.

(3) High water pressure

In deep mining, with the increase of in-situ stress, the water pressure of underground fissure rises. When the mining depth is greater than 1000 m, the water pressure will be as high as 7 MPa. With the increase of osmotic pressure, the effective stress of deep rock mass structure increases, driving the fissure expand, which leads to major engineering disasters such as water inrush of mine shaft.

1.3 Engineering response in deep mining

In the "three high" environment, the deep rock mass engineering responses present the characteristics of strong rheology, strong hot and humid environment, and strong dynamic disaster.

(1) Strong rheology

Deep rock mass has a strong time effect in the environment of high stress and high geotemperature, which is manifested as obvious rheology or creep. Under the interaction of high ground pressure, high geotemperature and high osmotic pressure, the hard granite is also prone to large-scale rheology.

(2) Strong hot and humid environment

Under the deep mining or excavation conditions, temperature of rock stratum will be as high as dozens of degrees centigrade, such as Russian kilometer average geotemperature of 30-40℃, up to 52℃ individually, a gold mine in South Africa, the geotemperature of 70℃ at 3000 m. The rise of geotemperature causes the reduced labor rate, and the underground workers are easily distracted or even unable to work.

(3) Strong dynamic disasters

In deep mining, the frequency and intensity of dynamic disasters are significantly increased. Rockburst, large area of roof weighting and inbrake, floor water inrush, impact ground pressure and so on are closely related to the mining depth.

The stress environment of deep mining is complex. Under the interacted impacts of mining effect, the mechanical characteristics of rock mass and its engineering response has obvious changes comparing with the shallow mining. At the same time, it causes rockburst, water inrush, large area of roof weighting and goaf instability and other catastrophic accidents intensify to a certain extent, such as severe accident frequency is increased, disaster mechanism is more complicated. It also brings huge challenges to the safe and efficient mining of deep resources. The way to solve these issues is the key to the safe and efficient mining of deep resources.

1.4 Status of deep metal mines at home and abroad

Overseas deep mining and researches have started earlier, in order to guarantee the smooth progress of deep mining, the governments, industrial sectors and research institutions of the United States, Canada, Australia, South Africa, Poland and other countries have cooperated closely to carry out the basic researches by closely combined with their human and financial resources and the relevant technologies of deep mining, and made the deep mining become an important field of mining researches. Since the 1980s, rockburst disasters have occurred frequently in the deep underground mines in Ontario, Canada. For this reason, the federal and provincial governments and mining industrial sectors have cooperated to carry out two 10-year deep mine research programs: The Canadian Ontario Industry Project (1985-1990) and the Canadian Rockburst Program (1990-1995). At the Creighton Mine of Sudbury, Canada, geophysical experts have conducted computer simulation researches of the statistical prediction of microseisms and rockburst at this mine. The Research Center for Rock Mechanics at the Laurentian University has carried out fruitful researches on supporting systems and rockburst risk assessment in potential rockburst zones. In the Idaho region of the United States, there are three production mines with a mining depth of 1650 m. Idaho University, Michigan Technical University and Southwest Research Institute have carried out deep mining researches on it, and cooperated with the U. S. Department of Defense to study the difference between seismic signals caused by rockburst and signals of natural earthquakes and nuclear explosions. Western Australia's mining industry has also faced challenges due to rockburst hazards in the deep mining, where the Center for Rock Mechanics at the University of Western Australia has done a lot of work. South Africa, which has the largest number of deep mines, launched the "Deep Mine" research program in July 1998, aiming to solve the key issues of safe and economic mining of deep gold mines at 3000-5000 m. The project costed 13.8 million US dollars and made a series of innovative achievements in deep mining technology of gold mines.

At present, there are 112 underground metal mines (deep mines) with the mining depth over 1000 m in the world. Ranking by quantity, the top five countries are Canada with 28, South Africa with 27, Australia with 11, the United States with 7, and Russia with 5. Among these 112 deep mines, 58 with the mining depth of 1000-1500 m, 25 with the mining depth of 1500-2000 m, 13 with the mining depth of 2000-2500 m and 16 with the mining depth over 3000 m (including 3000 m). More than 70% of them are gold mines and copper mines, and 12 out of the 16 mines with the mining depth over 3000 m are in South Africa, all of which are gold mines (Table 1-1).

Underground metal mines with the mining depth over 3000 m in the world   Table 1-1

No.	Name of mine	Mining depth(m)	Country
1	Mponeng Gold Mine	4350	South Africa
2	Savuka Gold Mine	4000	South Africa
3	TauTona Anglo Gold	3900	South Africa
4	Caritonville	3800	South Africa
5	East Rand Proprietary Mines	3585	South Africa
6	South Deep Gold Mine	3500	South Africa
7	Kloof Gold Mine	3500	South Africa
8	Driefontein Mine	3400	South Africa
9	Kusasalethu Mine Project, Far West Rand	3276	South Africa
10	Champion Reef	3260	India
11	President Steyn Gold Mine	3200	South Africa
12	Boksburg	3150	South Africa
13	LaRonde-mine	3120	Canada
14	Andina Copper Mine	3070	Chile
15	Moab Khotsong	3054	South Africa
16	Lucky Friday Mine	3000	The United States

China's underground metal mines account for about 90% of the total number of metal mines. With the gradual depletion of shallow resources, including some of the existing open pit mines will be moved into underground mining. The proportion of underground mining will be bigger, the mining depth will be deeper, the safe and efficient exploitation of deep energy and mineral resources is related to the major issues of the national economy's sustainable development and security to the national energy strategy. Since the 21st century, the most economic and effective safeguard measure for energy and mineral resources is the exploitation and utilization of deep resources in our country. In the next 10 to 15 years, one third of China's nonferrous metal mines will be mined to a depth of 1000 m or more. The maximum mining depth might reach 2000-3000 m, and the maximum mining scale might reach 20-30 million tons per year. In recent years, there are many metal mines in China with the mining depth of more than kilometers, such as Liaoning Hongtoushan Copper Mine with 1300 m, Jiapigou Gold Mine with 1400 m, Yunnan Huize Lead-Zinc Mine with 1500 m, Henan Lingbao Yinxin Gold Mine with 1600 m, Tongling Dongguashan Copper Mine with

1100 m, Shandong Sanshandao Gold Mine with 1050 m, Linglong Gold Mine with 1150 m, Xiangxi Gold Mine with 1100 m. In addition, Fankou Lead-Zinc Mine, Jinchuan Nickel Mine, Gaofeng Tin Mine and so on have entered the deep mining stage.

In recent years, a number of large and medium-sized metal mines under construction or planned to be built are basically deep underground mining. For example, Benxi Dataigou Iron Mine has 3 billion tons of ore reserves and with the mining depth of 1200-1600 m. With the ore body buried depth of 800-1600 m, Sishanling Iron Mine has 2.6 billion tons of ore reserves, and the first-stage mining scale of 15 million tons per year. Xianshan Iron Mine has 1.7 billion tons of ore reserves, and the mining scale will reach 30 million tons per year. A number of large gold deposits have been discovered in the deep below 2000 m in China. For example, a large gold deposit with 400 tons of gold reserves has been proved in the Xiling mining area of Sanshandao Gold Mine in Shandong Province at the depth of 1600-2600 m. It is predicted that there are still large metal deposits, especially nonferrous metal deposits and gold deposits, in the deep of 3000-4000 m in China. In particular, the implementation of early deep exploration and deep prospecting plan has made the distributions of mineral resources in the underground range of 3000-5000 m in the main ore-concentrated areas of China "transparent". According to the current resources mining status, the mining depth of metal mines in China is increasing at a rate of 8-12 m per year. In the next 10 to 20 years, China's metal mines will be mined at a depth of 1000-2000 m.

There are 16 underground metal mines in China with mining depth of 1000 m or more (Table 1-2).

The list of the underground metal mines with mining depth over 1000 m in China   Table 1-2

No.	Name of mine	Region	Mining depth(m)
1	Yinxin Gold Mine	Zhuyang Town, Lingbao City, Yunnan Province	1600
2	Huize Lead-Zinc Mine	Huize County, Qujing City, Yunnan Province	1500
3	Liuju Copper Mine	Liuju Town, Dayao County, Yunnan Province	1500
4	Jiapigou Gold Mine	Huadian City, Jilin Province	1500
5	Qinling Gold Mine	Guxian Town, Lingbao City, Henan Province	1400
6	Hongtoushan Copper Mine	Houtoushan Town, Fushun City, Liaoning Province	1300
7	Wenyu Gold Mine	Yuling Town, Lingbao City, Henan Province	1300

continued

No.	Name of mine	Region	Mining depth(m)
8	Tongguan Zhongjin	Tongyu Town, Tongguan County, Shaanxi Province	1200
9	Linglong Gold Mine	Linglong Town, Zhaoyuan City, Yantai, Shandong Province	1150
10	Dongguashan Copper Mine	Shizishan District, Tongling City, Anhui Province	1100
11	Xiangxi Gold Mine	Yuanling County, Huaihua City, Hunan Province	1100
12	Ashele Copper Mine	Ashele Area, Xinjiang Uygur Autonomous Region	1100
13	Sanshandao Gold Mine	Laizhou City, Shandong Province	1050
14	Jinchuan No. 2 Mining Area	Jinchang City, Gansu Province	1000
15	Shandong Jinzhou Mining Group	Rushan City, Weihai, Shandong Province	1000
16	Gongchangling Iron Mine	Gongchangling District, Liaoyang City, Liaoning Province	1000

1.5 Challenges in deep mining

After entering deep mining, the increase of in-situ stress, the deterioration of geological structure of deposits and occurrence conditions of ore bodies, the increase of broken rock mass, the increase of water inflow, the rise of mine temperature, and the serious deterioration of mining technical conditions and environmental conditions lead to the increases of difficulties in mining, the increase of disasters and accidents, the decrease of labor productivity and the sharp increase of costs. It brings a series of engineering and technical issues to large-scale normal production and safe and efficient mining of deep metal mines. The main difficulties to be faced and solved are as below:

(1) The rockburst, collapse, roof fall, water inrush and other mining dynamic disaster caused by the high in-situ stress field, which seriously threaten the safety of deep mining.

The in-situ stress increases linearly with depth. Rockburst is a process in which the disturbed energy caused by mining excavation accumulates and suddenly releases in rock mass. The higher the in-situ stress, the greater the mining disturbance energy, and the probability and magnitude of rockburst are higher. Due to the deep mining of a large number of gold mines, South Africa is the country with the most rockbursts in the world, with the maximum rockburst magnitude reaching $M_L 5.1$. From 1984 to 1993, 3275 miners died in mining accidents in gold mines in South Africa. The fundamental reason is the failure of studying and adopting proper mining methods conducive to rockburst control when mining below 2000 m.

It is relatively late for China's underground metal mines entering deep mining, few mines have entered into the deep mining in the last century. Thus few mines with rockbursts are observed, the time is late and the scale is not large. There are 8 underground metal mines in China that have experienced significant rockbursts up to now (Table 1-3). The rockburst occurred in Hongtoushan Copper Mine in the 1980s when it was mined to the depth of 400 m, and it gradually became more frequent after it was mined to the depth of 700 m. Rockbursts occurred almost every year, which have been mainly manifested by rock burst, tunnel wall caving, and roof inbrake, etc. In 1999, there were two large-scale rockbursts, the destructive forces of which were equivalent to 500-600 kg of explosive. In 2002, a rockburst caused the inbrake of tunnel roof, destroying mining equipment and forcing a halt to mining operations. In Dongguashan Copper Mine, rockburst occurred in 1997 when the mining depth was close to 1000 m, and in 1999, a significant rockburst occurred, which caused the destruction of a large number of rockbolt net support, and part of the rockbolt bar-mat reinforcement was cut off and thrown out as a whole.

The rockbursts in some metal mines in China   Table 1-3

Name of mine	Current mining depth(m)	Status of rockburst
Hongtoushan Copper Mine	1300	Strong rockburst occurred
Huize Lead-Zinc Mine	1500	Strong rockburst occurred
Dongguashan Copper Mine	1100	Moderate rockburst occurred
Lingbao Yinxin Gold Mine	1600	Moderate rockburst occurred
Erdaogou Gold Mine(Jiapigou Gold Mine)	1500	Moderate rockburst occurred
Linglong Gold Mine	1150	Moderate rockburst occurred
Sanshandao Gold Mine	1050	Minor rockburst occurred
Lingnan Gold Mine	800	Minor rockburst occurred

(2) The high temperature environment and heat damage control in deep mine mining.

The temperature of underground rock stratum increases with depth. According to statistics, below the normal temperature zone, the temperature of rock stratum will increase by about 1.7-3.0 ℃ for each 100 m of depth generally. Usually, the temperature of the rock stratum will exceed the human body temperature at the depth of more than kilometers. For example, the temperature of the rock stratum might reach up to 80℃ at the depth of 3000 m in the mines in western South Africa. At present, China has 16 underground metal mines with a mining depth of more than 1000 m, and more than 100 underground metal mines with a mining depth of more than 700 m. According to local statistics, for the mine shaft with the mining depth over 700 m, the temperature of the rock stratum is more than 35℃, some close to 40℃, and the highest reaches nearly 50℃. For example, in the Anhui Luohe Iron Mine, the measured temperature of the rock stratum is 38℃ in the east and 42℃ in the west at the depth of 700 m. Guangxi Hechi Gaofeng Tin Mine, such measured temperature reaches 40℃ at the depth of 700 m; Shandong Sanshandao Gold Mine reaches 38.5℃ at the depth of 825 m. Anhui Lujiang Nihe Iron Mine reaches 40.8℃ at the depth of 870 m. Such temperature values far exceed the stipulated standard in the China's *Safety Regulations for Metal and Nonmetal Mine* GB 16423—2020 " Works should be stopped if the wet bulb temperature exceeds 30℃". The high temperature leads to the serious deterioration of working conditions, which brings serious impacts to the safe operation of equipment, production efficiency, workers' health and labor productivity, etc. Because of the high humidity of underground working environment, the attention, judgment and coordination of underground workers will be reduced, which will affect the work efficiency of workers, and seriously lead to the occurrence of accidents. According to statistics, when the operating environmental temperature in the mine exceeds 1℃, the labor productivity of workers will be reduced by 7%-10%. Economic and effective measures shall be taken to solve the issues of high temperature environment and temperature reduction in deep mine, so that the working face of deep mine can maintain the temperature and humidity that the personnel and equipment can bear, thus ensure the normal work of deep underground mining.

(3) The issues on hoisting capacity and hoisting safety of deep mine mining.

Hoisting is as important as excavating in the mining process. With the increase of mining depth, the hoisting height is double increased. This has made production efficiency significantly reduced and production costs greatly increased, which also posed a serious threat to production safety.

The friction wheel multi-rope hoist is widely used in mines in China. In the range of depth less than 1000 m, this type of hoisting technology is most economical and efficient. Before 2000, China's underground mining depth of the vast majority is within 800 m, many of them are in the range of 500-600 m. In this depth range, the use of traditional friction wheel multi-rope hoist has been guaranteed to improve efficiency, cost, reliability, and safety. However, as the hoist height increases and the wire rope lengthens ceaselessly when entering deep mining, this hoist technology presents many difficulties in terms of hoisting capacity, safety, and operating costs. According to the statistical data of various countries, the friction wheel hoist will not be used after the mine depth is more than 1800 m. The main issue is that when the mine depth is more than 1800 m, the wire rope is lengthened and the hoisting load is increased. The weight of the wire rope might exceed the loaded weight of the hoisting container, so that the hoisting capacity is greatly reduced. When the wire rope is lengthened, its inertia increases greatly, which makes it difficult to control the stability of hoisting operation. After the wire rope is lengthened, the length of the tail rope changes significantly, leading to the wire rope's breaking and unevenness due to the large tension change. The effective metal section of the wire rope decreases, the tensile strength decreases, the life of the wire rope declines sharply, which has become the main factor limiting the safety and efficiency of the friction wheel hoist.

In order to overcome the deficiency of friction wheel hoist, Burrell from Britain has developed a multi-rope winding hoist. This multi-rope winding hoist solves the tail rope problem of the multi-rope friction hoist in the deep mine hoisting. It can be used for multilevel hoisting of double containers, and can be used for shaft excavation. Without tail rope, it is allowed to hang equipment and long materials at the bottom of the container. At present, this hoisting equipment is mainly used in South Africa, but not in other countries, and not in China.

(4) In order to cope with a series of difficulties in normal mining brought by deep mining high stress field, high mine temperature, high mine depth and complex geological conditions, improve mining efficiency, reduce production costs and ensure mining safety, it is necessary to carry out major changes to the traditional mining mode and mining methods.

After entering deep mining, as the in-situ stress increases, the geotemperature increases, geological structure and rock mass conditions will be changed greatly, the hard rock in the shallow mining might change into the soft rock in the deep mining, and elastic body might become plastic body or hidden plastic body. This will lead to serious deterioration of mining technical conditions and environmental conditions, increase the difficulties of mining, mining excavation, supporting, hoisting and transportation costs increase sharply. If do not change the traditional mining mode and mining methods, the normal production will be difficult to sustain.

The traditional mining mode and technological change involve many aspects. In addition to reforming the traditionalhoisting method, reducing the amount of hoisting is also an important aspect to solve the above-mentioned difficulties of deep mine hoisting. With traditional drilling and blasting mining method, adjacent waste rock is extracted along with ore, which is mixed up and hoisted out of the mine, thus increasing the hoisting capacity. Starting from the long-term goal, adopting continuous cutting equipment to replace the traditional blasting mining technology for rock breaking and mining is an important direction. Mining by the continuous cutting equipment is able to mine the target ore accurately, make the waste rock mixing rate reduce to a minimum, so as to greatly reduce the hoisting capacity. As blasting is not required for the mining space, it improves its stability. At the same time, the use of continuous cutting equipment for mining instead of the traditional blasting mining is also an inevitable need of implementing remote control intelligent mining and unmanned mine construction. It involves a large-scale transformation of traditional mining technology, and strategic changes of mining method and technology.

There are three types of mining methods in underground metal mines, namely, open-stope method, caving method and backfilling method. The costs of backfilling method is high. Except a few important nonferrous mines and gold mines adopting backfilling method, other mines, especially iron mines, generally adopt open-stope method and caving method for mining. After entering the deep mining over 1500 m, facing the high mining ground pressure, the open-stope method and caving method can not guarantee the safety of mining, and the backfilling method will be the mining method that must be adopted. This is also an important change to the traditional mining mode. Facing with this change, underground mines in China shall carry out systematic and innovative researches on various backfilling processes and backfilling materials, and form backfilling technology with low backfilling cost, high strength of backfilling body and wide source of backfilling materials, so as to create conditions for wide promotion of backfilling mining methods in deep mining.

(5) In order to better deal with the deteriorated deep mining conditions and environmental conditions, fundamentally ensure the safety of deep mining and improve mining efficiency, it is necessary to develop highly automated remote control intelligent unmanned mining technology.

The automatic intelligent mining utilizes advanced information and communication technology, remote sensing control technology, intelligent mining equipment, etc., to replace person with "robot" to complete various mining operations. It not only fundamentally resolves the threat to the mining safety of person, such as various disasters and accidents caused by the deep mine high temperature environment and the deep high stress, but also improves the mining efficiency and ore recovery to the greatest extent. With safety, all types of useful ores in complex and harsh conditions can be mined at deep depth.

As early as the early 1980s, Sweden, Canada, Finland and other western countries have begun the researches and site applications of remote control automation mining operations. The researches in this field in China have started at least 20 years later. It is not until the 11th Five-Year Plan period that national research projects appear, such as the "863" project at the 11th Five-Year Plan for *Development of Key Remote Control Technology and Equipment for Underground Mining (Unmanned Working Face)* and the "863" project at the 12th Five-Year Plan for *Intelligent Mining Technology for Underground Metal Mines*. Meanwhile, several metal mines represented by Shougang Group Xingshan Iron Mine, through joint study, research, and production, use for reference, introduce and adopt a batch of modern advanced technology, at the same time in the comprehensive advancement of digital mine construction, mainly through independent researches and development and integrated innovation, mine production automation and remote control of intelligent operation level also have great progresses. At present, automatic controls of the whole process of mining operations, including drilling, breaking, hoisting, belt transportation, drainage and ventilation, etc. have been basically achieved, but there is still little progress in this field nationwide. At present, many mines in China have not fully realized mechanical operation, so it is extremely difficult to popularize and apply remote control intelligent mining technology, and a lot of innovative researches shall be done according to China's national conditions.

(6) There are a large number of deep and medium-deep low grade metal deposits in China. With the increase of mining depth, the grade of metal deposits to the deep shows a decreasing trend. If the traditional mining method and technology are used to mine these deposits, the ore must be excavated from the deep and hoisted out of the mine first, and then the valuable metal elements in the ore shall be recovered through breaking and beneficiation. For low-grade ore and complex mining and beneficiation steps, the costs are extremely high and the economic benefit is extremely poor. Therefore, it is necessary to research and develop low-cost mining methods, such as in-situ breaking, leaching and biological recovery mining technologies, which do not need to excavate and hoist the ores, eliminating breaking and grinding steps. It will greatly reduce the costs. In-situ leaching and biological recovery mining technologies are just at the starting stages in China, the applicable types of metal mines are few. Thus, a lot of systematic development researches are required. In order to reduce the mining costs, it is necessary to study the methods and techniques to reduce the costs of ore hoisting and beneficiation, such as putting the main step of beneficiation under the mine.

2

Geological Conditions in Sanshandao Mining Area

The Sanshandao Mining Area is located at 25 km north of Laizhou City, Shandong Province, and adjacent to the newly built Laizhou Port. The administrative division is under the jurisdiction of Sanshandao Office. Geographic extremum coordinates: 119°56′59″-119°57′30″ east longitude; 37°24′00″-37°24′45″ northern latitude. There are railways, expressways and national highways around the mine. In the south, the mine is 20 km from Laizhou Station of Huang (Hua)-Yan (Tai) Railway, 30 km from the Laizhou south entrance of Wei (Hai)-Wu (Hai) Expressway. In the east, it is 16 km from Yan (Tai)-Wei (Fang) Expressway (National Highway 206), 26 km from Zhaoyuan and Sanshandao entrance of Wei (Hai)-Wu (Hai) Expressway. The newly built Laizhou Port is an important oil storage and transportation wharf. It is an open port with an annual throughput of 3 million tons. Both land and water transportation are convenient, as shown in Figure 2-1. In the mining area, the geological environment of the deposit is good, the engineering geological condition is medium complex, and the hydrological condition is complex.

The mining area is bordering the Bohai Sea in the west, with a low and flat topography with an average elevation of 2-3 m. There are three small hills near the coast, and the highest point is 67.14 m above sea level. It is a monsoon climate of warm temperate zone with four distinct seasons. The highest temperature is 38.9℃, the lowest is −18℃, and the average is 12.5℃ annually. It is dry in spring and winter, rainy in summer and autumn, with an average annual precipitation of 600-700 mm, the maximum annual precipitation of 1204.8 mm (in 1964) and the minimum of 313.8mm (in 1977).

The mining area is located in the east of Yishu fault zone. According to historical records, since 945 AD, nearly 100 earthquakes have occurred in Jiaodong region, among which 6 destructive earthquakes with magnitude above 6 occurred mostly to the east of Penglai and to the south of Jimo. Small earth-

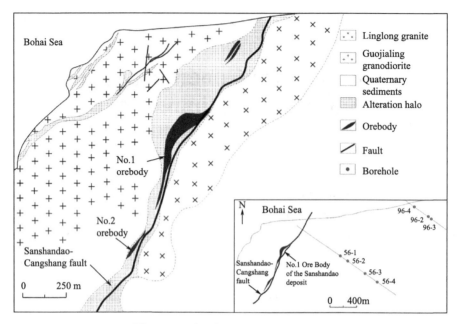

Figure 2-1 The geographical location of Sanshandao Gold Mine

quakes and microseisms below magnitude 6 occurred from time to time. There were two earthquakes that brought greater impacts to Laizhou. One was the Juxian and Tancheng earthquake, which occurred on February 25, 1668, with the magnitude of 8.5, intensity of 12 degrees, and the intensity of 8 degrees in Laizhou. The other one was the Bohai Bay earthquake occurred on July 19, 1969, with the magnitude of 7.4 and the intensity of 6 degrees in Laizhou.

2.1 Regional geology

The Sanshandao Gold Mine is located on the edge of the north China platform (Grade I), Jiaoliao Platform Uprise (Grade II) and Jiaobei Uplift (Grade III), adjacent to the west of Yishu fault zone (Grade II), and with a close proximity to the east of the Neoproterozoic Linglong Superunit intrusive rocks which is closely related to the mineralization of Sanshandao Gold Mine. The regions around the mining area include two important gold mine ore-controlling fractures, Sanshandao—Cangshang and Xincheng—Jiaojia, which are called Sanshandao—Jiaojia Gold Mine Field where the large and medium-sized gold deposits are intensively distributed.

The Sanshandao Mining Area is located in a small peninsula (Sanshandao) projecting to Laizhou Bay, surrounded by sea on three sides and land on the east. The main outcropped stratum in the mining area is the Cenozoic Quaternary, and the Paleoproterozoic Lugezhuang Formation of Jingshan Group is sporadically outcropped. The intrusive rocks in the area are widely developed, mainly including Neoarchean, Neoproterozoic and Mesozoic intrusive rocks.

The mining structure consisting largely of NE trending fractures is extremely developed, the largest is the Sanshandao—Cangshang fracture, the second is NW trending Sanshandao—Sanyuan fracture which belongs to the reactive fracture structure after the metallogenic period. The secondary fractures are mainly distributed in the middle and the footwall of the Sanshandao—Cangshang fault zone at NNE-NEE trending fractures. The secondary distributive fractures develop into several fault zones which constitute the main ore-controlling fracture structures in Sanshandao Mining Area.

There are four main ore-controlling fracture structures in the mining area, namely F1, F2, F3 and F4. Among which, F1 is located in the northeast side of NE trending to Sanshandao fracture, with deep development throughout the entire mining area. The fault zone with the width of 50-200 m is formed before mineralization, controlling No. 1 alteration zone and mineralization range. The occurrence of the main ore body is basically consistent with that of the ore body. The main fracture surface is located in the hanging wall of the ore body with 1-10 m, with good water-resistant, and groundwater in the hanging wall of F1 that is generally difficult to enter the pit. The fracture surface of F2 is outcropped and formed torsional fracture by the hanging wall moving northward and the footwall moving southward. The development characteristics of fissures on both sides are obviously different. The dislocation distance of pyrite quartz vein on both sides is up to 30 m, and it has good hydraulic conductivity. F3 is located at the northwest side of the NW trending Sanyuan—Chengjia fracture, with a large scale and depth of development, and the width increases with the increment of depth. It spans the whole Sanshandao Mining Area. Because it is formed in the late stage and the crushed zone is not cemented, it presents several lamprophyre veins and crushed zones. F4 is small in scale and develops only within the mining area. The characteristics of each fracture are shown in Table 2-1.

The characteristic of main fractures in the mining area   Table 2-1

No.	Strike (°)	Trend	Dip angle (°)	Length (km)	Depth (m)	Width of crushed zone (m)	Status of backfilling	Mechanical property	Hydrological property
F1	40	SE	35-41	>9	>600	>10	Well cemented	Pressure-torsion	Water-resistant
F2	10	NW	85	≈0.5	—	—	—	Torsion	Hydraulic conductivity
F3	300	NE/SE	80-90	≈70	>600	17~36	Backfilled with breccia	Tensional and torsional	Hydraulic conductivity
F4	15	SE	40-45	≈0.3	≈300	≈1	Backfilled with mud	Pressure-torsion	Water-resistant

2.2 Engineering geology

2.2.1 Stratum conditions

The outcropped stratum in the Sanshandao Mining Area is mainly the Quaternary, followed by the inclusion of Jiaodong rock group at the drilling in the Lujia unit on the hanging wall of Sanshandao-Cangshang fracture.

(1) Quaternary

It is mainly composed of coarse, medium and fine sand and sludge deposited in the Xukou Formation, with the thickness of 30-40 m and the maximum thickness of 50 m. The lithology of this formation consists of four lithologic layers from bottom to top:

1) Sandy clay, clay layer: mainly yellow and brown clay containing gravel clay and reddish brown clay, the thickness is mostly 3-5 m. It distributes on the weathered crust of bedrock, with the continuous distribution, which has good water-resistant performance.

2) Medium coarse sand gravel layer: medium sand, coarse sand and gravel account for the majority. The thickness fluctuates greatly and the distribution is discontinuous. The thickness is generally 3-4 m, and the maximum thickness is up to 12 m, with good water abundance.

3) Sandy clay layer: mainly sandy clay, followed by sandy clay containing calcareous nodules and clayey sand, partial inclusion of sand and gravel lens. The thickness is generally 7-8 m, with a certain degree of water-resistant, but relatively poor.

4) Medium coarse gravel layer: mainly yellow brown medium, coarse sand, and the local contains fine sand and gravel. Organic matter, shell and black mud content increase downward. Thickness ranges from 3 m to 17 m, and the average thickness is 10 m. The water abundance varies greatly due to the amount of mud content.

(2) Inclusion of Jiaodong rock group

According to the prospecting engineering, the inclusion of Jiaodong rock group is Guogezhuang Rock Formation. Its lithology is mainly biotite plagiogneiss and biotite granulite, followed by biotite schist.

Biotite granulite: dark gray, fine grained microscopic flaky granular metacryst structure, with dense massive structure. The main minerals are feldspar, quartz and biotite. The feldspar is mainly alkaline feldspar, biotite evergreen chlorite, mostly flaky granular polycrystalline, and quartz is amorphous fine-granular crystal.

Biotite plagiogneiss: grayish green, flaky granular metacryst structure, with gneissic structure. The main minerals are plagioclase (45%), quartz (30%), biotite (15%) and amphibole (10%). The accessory minerals include apatite, sphene and phosphoite. Plagioclase is andesine and oligoclase, with hypautomorphic—xenomorphic granular crystal. Quartz is xenomorphic granular crystal, undulate extinction, directionally elongated. The biotite is a hypautomorphic flaky crystal with directional arrangement. Amphibole is an idiomorphic—hypautomorphic granular crystal. The accessory minerals are idiomorphic granular crystals.

Biotite schistt: grayish black, medium-fine-coarse-grained flaky metacrystal structure, with flaky structure, and the mineral compositions are mainly biotite, plagioclase, quartz and a small amount of garnet.

2.2.2 Status of structure

The structure of the Sanshandao Mining Area is dominated by fractures. The largest fracture is NE trending Sanshandao—Cangshang fracture, followed by NW trending Sanshandao—Sanyuan fracture. The secondary fractures are mainly distributed in the middle and the footwall of NNE-NEE trending fractures of Sanshandao—Cangshang fault zone. In the vicinity of the fracture, because the stress release is not easy to accumulate the elastic potential energy, the possibility of rockburst is low. However, if the direction of roadway layout is consistent with or similar to the fracture strike within a certain distance, the rockburst phenomenon is easy to occur on the cave wall of the adjacent fault zone.

(1) **Sanshandao—Cangshang fracture**

This fracture is outcropped in the north section of Sanshandao Mining Area. The surface outcrop and engineering control length is 1700 m, and the width of the tectonic rock zone is 50-200 m. The overall strike is 35°, with the trend of southeast, the dip angle range is 35°-45°, both the strike and the trend are soothing wavy, with the apparent pressure torsion. It is composed of fault gouge, mylonite, structural breccia, sericite, sericitization, silicification, sericitization cataclastic rock and altered cataclastic monzonite, which is the main ore-guiding and ore-controlling structure in the mining area. The development of gold ore body in Sanshandao Mining Area is in this tectonic alteration zone. At present, the maximum tilt extension of the fault zone controlled by the project is 1450 m, not only has not been pinched out, but also the thickness is still more than 30 m. The fracture has been active for many times, and there are several structural planes in the zone. Some of the structural planes are mine developed and have good continuity, with certain water-resistant.

(2) **Sanshandao—Sanyuan fracture**

This fracture (called F3 in the mining area) is located at the northwest side of the Sanshandao—Sanyuan fracture in the region, with the length of 1500 m in the mining area. It extends into Laizhou Bay in the northwest direction and extends beyond the mining area in the southeast direction. The fault runs through the whole mining area with the depth of more than 600 m. The strike of the fault is 290°-300°, the trend is mainly NE, partial anti-dip, with the dip angle over 80°. The fault structure crushed zone is 10-25 m in width, which is composed of several backfilled lamprophyre and other basic vein rocks, cataclastic rocks and breccia. The width of the basic vein rock is 0.3-3 m, with the majority of 0.7-1.3 m. The rock mass is relatively broken, with montmorillonite, and the width of breccia zone is 0.3-1.5 m. There are gray fault gouge in the shallow part and gradually decrease in the deep direction.

This fracture has two active records, one of which is extensional, resulting in the intrusion of a large number of basic vein rocks along the fault. The second is left translation, the result of activity is lamprophyre breaking, with stagger left horizontal. According to the characteristics of xenoliths or breccias of pyrite sericite contained in the lamprophyre in F3 fracture, it can be concluded that this fracture is a highly active after mineralization.

Due to the multiple activities of this fracture, especially after the mineralization period, the destruction of the ore body in the mining area is relatively obvious. According to the actual data of mine production, the fracture horizontally dissects the main ore body in the mining area by about 20 m. The fracture not only breaks itself, but also promotes the development of rock fissures on both sides, which greatly reduces the stability of rock mass and brings some difficulties to the mining of the deposit. This crushed zone is dominated by breccia, which is not cemented, with good water abundance and hydraulic conductivity. Because it cuts the Sanshandao—Cangshang fracture and extends into the Bohai Sea at the northwest side, it not only partially destroys the aquiclude of the Sanshandao—Cangshang fracture, but also communicates the hydraulic connection between the fractures in the mining area, resulting in a large amount of water gushing in the mining tunnel.

(3) Secondary fracture

The secondary fracture refers to the NNE-NEE trending fracture structure occurred in the middle and the footwall wall of Sanshandao fracture.

1) NEE trending fracture (F1)

This fracture is developed in the Sanshandao section of Sanshandao—Cangshang fault zone. It is located in the middle and upper part of the fault zone at Sanshandao, with the width of 6-40 cm, and consists of fault gouge, mylonite and structural breccia. The fault strike and trend are relieved wavy with obvious pressure torsion. The main section extends steadily, and there are cataclastic rock belts of 1-10 m thickness on both sides. This fault is the main ore-controlling structure during the metallogenic period, and the fault gouge stably distributed along the main section plays a blocking and enrichment role for the rising mineralization hydrothermal fluid in the deep. Therefore, the main ore body of the gold mine occurs below the main fracture surface of the fault. After mineralization, the fault activity is weak and extensional, which leads to the partial fragmentation of the ore body.

2) NNE trending fracture (F2)

It is located at about 200 m to the west side (footwall) of Sanshandao—Cangshang fracture. The strike is 12°, the trend is NW and the dip angle is 85°. The fault has the length of 600 m and extends northward into the Bohai Sea. It is a sinistral torsional fault with a crosscut pyrite quartz vein and a dislocation of 20 m. No fault gouge, mylonite and breccia are found on the surface of the fault. The drilling data shows that it disappears when it extends to 250 m.

3) NEE trending fracture

It is located at about 300 m to the west side (footwall) of Sanshandao—Cangshang fracture. A distributive fracture with a length of 900 m has developed, with the overall trending of 50° and a gentle wavy shape. Due to the lack of deep engineering control, its trend and dip angle have not been proved. The tectonic rock zone is 5-20 m in width, with tectonic alteration zone backfilled by sericitization, silicification, sericitization hydrothermal alteration and quartz veins.

2.2.3 The distribution of lithology

The ore body of Sanshandao Gold Mine is pyrite sericite, and the main rocks are porphyritic biotite granite, pyrite sericite granite and pyrite sericite. The surrounding rock of the hanging wall and footwall of the ore body is sericite, and the indirect surrounding rocks are biotite hornblende tonalite or biotite plagiogneiss, and biotite granulite, etc. The rocks of footwall are relatively complete, and the rocks of hanging wall are relatively broken. There are three rock formations in the mining area, which can be divided into loose weak rock formation, weathered and tectonic alteration rock formation and block rock formation according to lithology and engineering geological condition.

(1) The loose weak rock formation is the Quaternary accumulation, covering the bedrock. The lithology is medium coarse sand, gravelly sand, silty-fine sand and sandy clay. The thickness is 30-50 m, with abundant groundwater. This rock formation is mainly composed of loose sand and cohesive soil. It belongs to hard plastic rock with low mechanical strength and poor engineering geological conditions.

(2) The weathered thickness of the weathered and tectonic alteration rock formation bedrock is generally 10-20 m. Weathered rock is a weak rock with poor engineering geological condition due to the development of rock fissures with poor integrity and generally less than 20 MPa compressive strength. The tectonic

alteration rock has developed rock fractures and with poor integrity. The primary rock structure has been changed due to the alteration. The compressive strength is 14.9-50.8 MPa, and it belongs to weak—semi-hard rock with poor—good engineering geological conditions.

(3) The block rock formation is widely distributed in the mining area, the lithology is monzonite and biotite hornblende tonalite, with block structure and complete rock mass. The compressive strength is greater than 60 MPa, it is a hard rock, with good engineering geological condition.

Large weak structural planes such as F1 and F3 faults are the main factors affecting the stability of deep rock mass, and deep engineering geological issues are closely related to the existences of such weak structural planes. The tunnel is located in the footwall of F1 fracture, and the NW trending structure is relatively developed. The rocks in the fault zone and its vicinity are crushed, and it is easy to produce falling blocks and collapses during excavation. The F3 fracture strikes NW with a dip angle of nearly 90°. The rocks in the fault zone are broken, mainly backfilled with gravel, mudstone, kaolin clay, sand gravel and clay. Long-term immersion by groundwater makes it in a state of water saturation, which leads to the reduction of frictions among the filled particles. When the tunnel is excavated to the F3 fracture, it is easy to cause the tunnel debris flow disaster.

2.3 Hydrogeology

2.3.1 Geomorphology, hydrology and meteorology

The topography on the region is high in the southeast and low in the northwest. The southeast is a low-lying hilly area composed of granite of Yanshan period and Jingshan Group stratum, with great topographic relief and generally 50-90 m ground level. Wangershan Mountain is the highest point in the region, with the level of 177.39 m. The northwest topography is flat, with the alluvial-proluvial and marine depositional plains of Wang River and Zhuqiao River, with the ground level of 2-50 m. The main developments in the region are denudation accumulation and accumulation landform types.

This region is a warm monsoon continental subhumid climate with four distinct seasons. According to the statistical data of Laizhou Meteorological Bureau for many years, the annual average temperature is 12.5℃, the extreme minimum temperature is −18.0℃, and the extreme maximum temperature is 38.9℃. The average annual precipitation is 612.1mm, the maximum annual precipitation is 1204.8 mm, and the minimum annual precipitation is 313.8 mm. The precipitations are mostly concentrated in July to September, accounting for more than 60% of the annual precipitations. The annual average evaporation is 1665.1 mm, the average relative humidity is 64%, the maximum snow depth is 40 cm, the maximum permafrost depth is 68 cm, and the once-in-a-century tsunami invasion level is 3.95 m.

There are two main water systems in the region, i.e. Wang River and Zhuqiao River. The Wang River originates from the Daze Mountain in the southeast, with the total length of 48 km and a drainage area of 376 km^2, and flows into the Bohai Sea through the south side of the mining area. It is an intermittent river with a longer dry period and no more than 10 days with streaming flows in the summer. Zhuqiao River is located at 8 km in the east of the mining area. It originates from the eastern hills and flows into Laizhou Bay from south to north. It is a seasonal river.

2.3.2 Hydrogeological conditions

The mining area is located in the runoff excretion area of groundwater of regional hydrogeological unit. The topography is flat with wide coverage of Quaternary in the region, with the ground level of generally 1-6 m. There are three connected hills in the northwest, with the highest point level of 67.1 m; The mining area is connected with the land in the southeast, Laizhou Bay is at the northwest, and the sea level of the Bohai Sea is the local lowest erosion base level.

(1) Quaternary pore aquifer and aquiclude

The Quaternary is widely distributed in the mining area. According to the lithology, it is generally divided into four layers from top to bottom. The first and third layers are aquifers, and the second and fourth layers are relative aquiclude or aquiclude.

The first aquifer: it is mainly composed of medium and coarse sand, with fine sand and gravel in partial areas. The thickness is 3.50-17.29 m, the unit-water-inflow of drilling is 0.31-15.27 L/s · m, the permeability coefficient is 5.35-117.46 m/d, and the difference of water abundance is great. This aquifer mainly accepts meteoric precipitation and seawater supply. The buried depth of water level is 0.5-6 m. The water quality varies greatly, and the salinity is 0.21-25.95 g/L, which increases from east to west and south to north. With the increase of salinity, the hydrochemical types change from Cl • HCO_3—Na • Ca type to Cl—Na • Mg type and Cl—Na type.

The first aquiclude: it is located under the first aquifer with the buried depth of 5.5-9 m. The lithology is mainly sandy clay, and sandy clay containing calcareous nodules and clayey sand, etc., with water-bearing lens of sand and gravel interbedded locally. The thickness changes little, generally 7-8 m. The viscosity of this layer is low, with relatively poor water-resistance.

The second aquifer: it is located under the first aquiclude, which is discontinuous and mainly distributed to the east of Line 32, south of Hole ZK7 of Line 40 and the southwest of Hole ZK16 of Line 20. The lithology is mainly medium coarse sand and gravel. The thickness gradually increases from north to south, generally 3-4 m, with a maximum of 11.9 m. It contains pore confined groundwater, and can accept the first aquifer and seawater supply.

The second aquiclude is located above the bedrock weathered crust with the buried depth of 7.8-25.5 m. This layer is stable and only missing in Hole ZK56 and observation Hole 5 of Line 64 in the whole mining area. The lithology is mainly yellow brown sandy clay with gravel and reddish brown clay. The thickness is generally 3-5 m, the maximum thickness is 19.6 m, with high viscosity and good water-resistance.

(2) The water-bearing zone of the F1 hanging wall fissure (II)

There are weathered fissures and structural fissures in the granite on the hanging wall of F1, which contain weakly confined groundwater. According to the drilling data, the shape of the water-bearing zone is extremely irregular, with a general dip angle of 10°-45° to the southeast. The roof is granite and metamorphic rock, and the westernmost part is the Quaternary clay layer. The unit-water-inflow of drilling is 0.001-0.041 L/s • m, with the salinity of 39.1-92.7 g/L.

The roof and floor of this water-bearing zone are porphyritic monzonite, medium-grained monzonite and metamorphic rocks of Jiaodong Group. The floor is partly pyrite sericite or a small amount of pyrite sericite granite. The rocks are dense and hard and relatively water-resistant. The F1 fault gouge and mylonite in the lower part have good water-resistance for the groundwater in the water-bearing zone against entering the pit.

(3) **The water-bearing zone of F3 fracture structure (I)**

F3 fracture is the NW side of the regional major fracture of Sanshandao—Sanyuan across the mining area. It is located in No. 32 to No. 36 prospecting line (near the surface) in the mining area, between 300°-310°, with overall NE trending, crosscutting F1 fracture and No. I alteration zone and the ore body therein.

The F3 fault zone has the developed width of 15-36 m and is characterized by multi-phase activities. Due to the large scale of F3, the broken zone formed by the later activities is not mine cemented. Therefore, there are abundant underground water stored and strong water conduction capacity. The long-term immersions of fault gouge and crushed zone by the groundwater cause its mechanical properties become poor. Almost each middle section has occurred water inrush, debris flow, and collapse etc. when passing through the F3 fault. In addition, the cutting dislocation of F3 destroys the water-resistance of F1 in local, thus makes it become the water-conductive structure between south and east regional groundwater. Although its northwest side is in the sea, it has not imported a large number of seawater due to good sealing condition.

According to the deep geophysical exploration, it shows that the F3 fault zone still exists obviously at the level of -600 m, indicating that the developed depth of F3 is large. The derived brine temperature has a tendency to increase with the deepening of development. Moreover, the supply source is sufficient, also indicating that the developed depth of F3 is large.

The water level in the F3 fault zone decreases with the continuous water inrush in the process of deep development of mine tunnel. The water quality of groundwater is complex, and the salinity at the initial stage of exposure is as high as 50-60 g/L. After one to several years of excretions, the salinity of the majority is gradually close to that of seawater, and the salinity of some water droplets with small flow can be lower than that of seawater. Water temperature increases with depth.

(4) **The water-bearing zone of the F1 footwall structural fissure (Ⅲ)**

There are several formations of structural fissures developed in the footwall of F1, which become the main space for groundwater activity. The water-bearing zone of the F1 footwall structural fissure is widely distributed, reaching Line 28 in the south and the primary No. ⑤ and No. ⑥ ore bodies in the north. F1 is the boundary in the east and gradually terminates in F2 in the west. The characteristics of structure development and the supplying conditions of groundwater are different in different regions, and the hydrogeological characteristics are also different. According to its hydrogeological characteristics, it is divided into two regions.

The structural fissure water-rich area of the F1 football (Ⅲ-1): it mainly distributes in F3 to Line 46, including distribution range of LD1-LD6, NW trending water-conducted fissure development, with larger fissure scale and larger extension. With stronger aquosity, unit-water-inflow of drilling is 1-2 L/s · m, and the maximum water inflow of water gushing point in the pit is 194 m^3/h. In the northwest, it receives the overflowing seawater supply, and in the southeast, it communicates with the water-bearing zone of F3 fracture structure and receives the regional brine supply. The area has good hydraulic connection, but it is easy to dry.

The water-bearing area of the F1 footwall structural fissure (Ⅲ-2, Ⅲ-3):

Ⅲ-2: it distributes in the north of Line 46 and the primary No. ⑤, No. ⑥ ore bodies. With the gradual decreases on density and scale of water-bearing fissure away from F3 and NW trending, the water abundance of water-bearing zone obviously weakens. The maximum unit-water-inflow of drilling is 0.224 L/s · m, and the maximum water inrush at the exposure of the tunnel is 50 m^3/h. The northwestern part receives the overflowing seawater supply and a small amount of penetration supply of the quaternary pore water. The water quality at the water gushing point in the region is still saltier than seawater at the beginning, while due to the proximity to the sea, the water desalinates to seawater quickly, and the water temperature is 18-23 ℃.

Ⅲ-3: it distributes at Line 28 to F3, the water-bearing structure mainly with several NW trending fissures, with moderate scale. The water-bearing structure adjacent to the south of F3 is backfilled with mud content, the strike extends from dozens of meters to hundreds of meters, and communicates with NE trending fissure. The maximum water inflow at a single water gushing point in the mine tunnel is 300 m^3/h, and the underground water level decreases with the decrement of the water gushing elevation of the mine tunnel. The water quality at the water gushing point is initially brine with high salinity, and shows desalination trend with the extension of drainage time, but with slower desalination speed. The water temperature is generally 26-37.5 ℃. The farther away from the F3 fault zone, the lower the water temperature is.

(5) Water-resistant rock mass

It distributes in the region south of Line 28, north of F3 and west of F2, and the fractures are not developed in this region. The unit-water-inflow of drilling is 0.00037-0.011 L/s · m. According to the geophysical prospecting, there is no obvious water-bearing channel to the south of Line 28, and no water gushing phenomenon at the exposure of tunnel. There are only four NE trending fissures with smaller-scale to the west of F2 and north of F3 range. The water level of observation hole in this region is over 100 m than that in the east of F2. Therefore, the two granite can be viewed as relatively water-resistant rock mass or water-resistant rock mass.

2.3.3 Groundwater type and mine water supply

According to the occurrence conditions, hydrological properties and hydraulic characteristics of groundwater in the region, it can be divided into two types: (1) pore water of loose rock type; (2) fissure water of block rock type.

(1) Pore water of loose rock type

1) Slope diluvial (QD) pore water: it mainly distributes in the Chengzi area in the eastern part of the region. The lithology is argillaceous sand and gravel, the thickness is about 8 m, the buried depth of water level is 1.8-3.7 m, the single mine water inflow is generally less than 500 m^3/d. It is submersible, with good water quality, the salinity is less than 1 g/L, and the hydrochemical type is mainly HCO_3—Ca type.

2) Alluvial and alluvial-proluvial (QY, QL) pore water: it distributes along the river and on its both sides, it is formed by alluvial and alluvial-proluvial. The lithology is medium coarse sand and gravel in the riverbed section. The two sides of the river have a dual structure, the upper part is clayey sand, the lower part is medium coarse sand and gravel, and the partial interbedded sandy clay lens, with the thickness of 5-30 m. It is submersible, with local micro-pressure, single mine water inflow of 500-1000 m^3/d, with good water quality. The salinity is less than 1 g/L, hydrochemical type is mainly HCO_3—Ca type.

3) Marine depositional (QXK) pore water: it distributes in the northern and western coastal areas. The lithology is argillaceous silty-fine sand, medium sand, gravel and pebbles at the bottom, and sandy clay lens with the thickness of 20-40 m and the maximum thickness of 60 m. The buried depth of water level

is 0.2-3.5 m, and the water inflow of a single mine is generally less than 500 m^3/d. The water quality is poor, and the highest salinity is 153.7 g/L. The hydrochemical type is mainly Cl—Na type.

(2) Fissure water of block rock type

It mainly distributes in the eastern hilly area, and its lithology is the granite and granodiorite, etc. in the early period and ahead of Yanshan period. Groundwater mainly occurs in the weathered fissures and structural fissures. The thickness of weathered zone is generally 20 m, and the water inflow of a single mine is generally less than 100 m^3/d. Under the influence of the structure, the fissures develop in the local area and the water abundance is enhanced.

The supply, runoff and excretion conditions of groundwater are affected by topography, geomorphology, lithology, meteorology, hydrology, geology and other factors, and the movement direction is basically consistent with surface water system.

Meteoric precipitation is the main supply source of groundwater in the region. In hilly area, most of the precipitation is converted into surface runoff and a small part penetrates into the ground to supply groundwater due to the larger topographic relief and deeper cutting. Groundwater in the plain area not only receives the supply of meteoric precipitation, but also receives the penetration supply of the bedrock fissure water and surface water, with good supply conditions. The flow direction of groundwater is basically the same as that of surface water, that is, it moves from southeast to northwest, and the runoff is excreted in Bohai Sea. The main excretion way is runoff, and some of them are excreted in the form of spring water in the low-lying topography, and artificial mining is also an important excretion way. The alluvial-proluvial pore water is the main source of the industrial, agricultural and domestic water in this region. In addition to receiving meteoric precipitation on-site supply, tidal seawater is also the main supply source for marine depositional pore water, and evaporation is its main excretion pathway.

3

In-situ Stress Measurement Method and Its Application

3.1 The principle of in-situ digitizing hollow inclusion cell

3.1.1 The in-situ digitizing hollow inclusion cell

The hollow inclusion cell has been successfully developed by R. Walton and G. Wolotnicki from the Commonwealth Scientific and Industrial Research Organization (CSIRO) in Australia in the early 1970s. Its main body is a hollow cylinder with the wall thickness of 3 mm made of epoxy resin, with an outer diameter of 37 mm and an inner diameter of 31 mm. Three groups of strain rosettes are embedded in the middle layer with 35 mm in diameter, along the same circumference at the same distance. Each group of strain rosette is composed of three foil gauges with an interval of 45°. The outstanding advantage of the hollow inclusion cell is that the strain gauge and the hole wall are cemented together in a large area, and the cementing agent can also be injected into the fissures and defects of the surrounding rock mass to make the rock whole. Therefore, the cementing quality is better and the complete trepan boring rock core is easy to be obtained. At present, it has become the most widely used in-situ stress relief measuring instrument.

Professor Cai Meifeng has made the significant improvements on the traditional hollow inclusion cell measurement technology, and developed to realize the complete temperature compensation and considering rock mass nonlinear in-situ stress relief hollow inclusion measurement technology and equipment, improved in-situ digitizing hollow inclusion cell (Figure 3-1), as well as realized miniaturization of in-situ stress data collection system, and wireless data transmission (Figure 3-2).

The characteristics of in-situ digitizing hollow inclusion cell:

3 In-situ Stress Measurement Method and Its Application

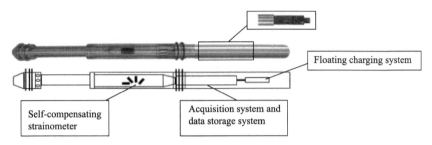

Figure 3-1 The in-situ digitizing hollow inclusion cell

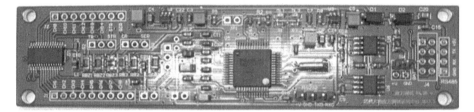

Figure 3-2 The micro-collection card of the in-situ digitizing hollow inclusion cell

① It can be used as a stress monitor for short-term or long-term stress monitoring;

② It adopts in situ digitizing design, without wire lead;

③ The circuit of the micro-collector adopts ADuC847 microprocessor chip, and is powered by 5V storage battery. The stabilization chip replaces the traditional stabilization module. The multi-channel simulation switch controls the opening and closing of each channel respectively, with the function of drift self-compensation;

④ It compensates and revises the temperature change of the collection channel and the collector itself based on the complete temperature compensation principle;

⑤ It has the power-off continuous collection function, with the accuracy of $+20$ $\mu\varepsilon$ under the same temperature condition;

⑥ It adopts high-strength non-magnetic aluminum alloy material, and the instrument cavity at the back of the device is encapsulated with heat-conducting silica gel, which has good waterproof, heat dissipation and shock absorption performances;

⑦ There are four working modes for the collector by continuous working, regular working, micro-standby working and standby working. The working modes can be customized according to different requirements of in-situ stress collection or stress monitoring.

⑧ The collecting interval is adjustable from 1 to 240 min, and the stored data can be extracted at any time according to the record number. The current operation parameter instruction can be viewed at any time, and it has the function of unlimited transmission.

The core technology of in-situ digitizing hollow inclusion cell:

① The miniaturization of collection system

On the premise of ensuring the accuracy and stability of collection, the collection plate circuit with the functions of instantaneous collection, power-off continuous collection, and drift self-compensation, etc. is developed. The stabilization chip replaces the traditional stabilization module and the multi-channel simulation switch controls the opening and closing of each channel respectively. The collection circuit has 16 input channels, of which channels 1-12 are strain signal channels, channels 13 and 14 are dual-temperature compensation channels, and channels 15 and 16 are drift compensation channels within the circuit.

② The electric bridge instantaneous collection technology

The signal collection is completed instantaneously when the electric bridge circuit is connected. There is no heating influence on the resistance and resistor disc for the instantaneously switching on, and no need of long-term power supply. It can be realized to collect upon switching on power immediately, and the signal is the true resistance value of the resistance foil gauge itself.

③ The technology of power-off continuous collection

It solves the problem that the traditional collector needs to be leveled again after power off and can not continue collection, and realizes the relevance and validity of the continuous data collection after power-off.

④ The measurement technology of wide range simultaneous temperature channel

With real-time monitoring of ambient temperature, it studies and develops strain channel with wide range, which can realize the strain measurement and synchronously access to thermal couple in the mean time. The channel temperature data will be obtained when temperature resistance value changes within $+100\ \Omega$.

⑤ The dual-temperature compensation technology based on complete temperature compensation principle

a. With the temperature compensation of measuring circuit, the temperature self-compensation foil gauge is used to reduce the temperature sensitivity of measuring channel. The temperature self-compensation design and temperature calibration test shall be carried out according to the field lithology in order to control the temperature coefficient of the measurement channel within 5 microstrain/°C.

b. With the temperature compensation of the collection circuit, temperature calibration (2 ppm) is carried out by the resistance of the low-temperature system. The temperature change of the collection circuit is measured simultaneously by the simultaneous dual-temperature channel. In the later temperature calibration test, the algorithm compensates to reduce the influence of temperature in the circuit.

3.1.2 Calculation method of in-situ stress component

As shown in Figure 3-3, the clockwise order of strain rosettes from the head to tail of the strain gauge are A-B-C. The foil gauge of parallel drilling strike is No. 0 foil gauge, which is 45°, 90° and 135° in clockwise order. Taking group A foil gauge as an example, No. 1 foil gauge is ring foil gauge, and denoted as A_{90}; No. 2 foil gauge is axial foil gauge, denoted as A_0; No. 3 foil gauge and No. 4 foil gauge are ±45° with the drilling axis respectively, denoted as A_{45} and A_{135}. Similarly, No. 5-No. 12 foil gauges are denoted as B_{90}, B_0, B_{45}, B_{135}, C_{90}, C_0, C_{45} and C_{135} respectively.

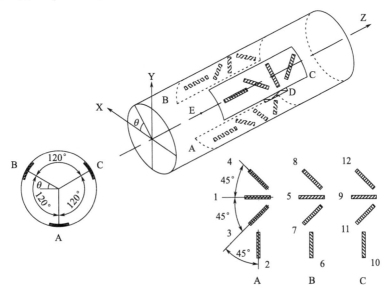

Figure 3-3   The improved type of hollow inclusion cell

By using the measured strain value of hollow inclusion, the primary rock stress value can be calculated according to the simultaneous equations of (3-1)-(3-4).

$$\varepsilon_\theta = \frac{1}{E}\{(\sigma_x+\sigma_y)k_1 + 2(1-v^2)[(\sigma_y-\sigma_x)\cos2\theta - 2\tau_{xy}\sin2\theta]k_2 - v\sigma_z k_4\} \tag{3-1}$$

$$\varepsilon_z = \frac{1}{E}[\sigma_z - v(\sigma_x+\sigma_y)] \tag{3-2}$$

$$\gamma_{\theta z} = \frac{4}{E}(1+v)(\tau_{yz}\cos\theta - \tau_{zx}\sin\theta)k_3 \tag{3-3}$$

$$\varepsilon_{\pm 45°} = \frac{1}{2}(\varepsilon_\theta + \varepsilon_z \pm \gamma_{\theta z}) \tag{3-4}$$

where σ_x, σ_y, σ_z, τ_{xy}, τ_{yz}, τ_{zx}—the three-dimensional stress components at the measuring point;

ε_θ—the circumferential strain value of the hole wall;

ε_z—the axial strain value of hole wall;

$\varepsilon_{\pm 45°}$—the strain value in the direction of $\pm 45°$ with the drilling axis;

$\gamma_{\theta z}$—the shear strain value;

E—the rock elastic modulus at measuring point;

v—Poisson's ratio of rock at measuring point;

θ—the included angle between foil gauge and the x axis, taken the counterclockwise rotation count as positive;

k_1, k_2, k_3, k_4—the four revision coefficients given by G. Worotnicki and R. Walton, collectively referred to as K coefficients, which can be calculated by equation of (3-5).

$$\left.\begin{array}{l} k_1 = d_1(1-v_1 v_2)\left(1 - 2v_1 + \dfrac{R_1^2}{\rho^2}\right) + v_1 v_2 \\[6pt] k_2 = (1-v_1)d_2\rho^2 + d_3 + v_1 \dfrac{d_4}{\rho^2} + \dfrac{d_5}{\rho^4} \\[6pt] k_3 = d_6\left(1 + \dfrac{R_1^2}{\rho^2}\right) \\[6pt] k_4 = d_1(v_2-v_1)\left(1 - 2v_1 + \dfrac{R_1^2}{\rho^2}\right)v_2 + \dfrac{v_1}{v_2} \end{array}\right\} \tag{3-5}$$

$$\left.\begin{aligned}
d_1 &= \frac{1}{1-2v_1+m^2+n(1-m^2)} \\
d_2 &= \frac{12(1-n)m^2(1-m^2)}{R_2^2 D} \\
d_3 &= \frac{1}{D}[m^4(4m^2-3)(1-n)+x_1+n] \\
d_4 &= \frac{-4R_1^2}{D}[m^6(1-n)+x_1+n] \\
d_5 &= \frac{3R_1^4}{D}[m^4(1-n)+x_1+n] \\
d_6 &= \frac{1}{1+m^2+n(1-m^2)} \\
D &= (1+x_2 n)[x_1+n+(1-n)(3m^2-6m^4+4m^6)] \\
&\quad +(x_1-x_2 n)m^2[(1-n)m^6+(x_1+n)]
\end{aligned}\right\} \quad (3\text{-}5)$$

where R_1—radius in the hollow inclusion;
R_2—radius of the installed small hole;
G_1—stiff modulus of hollow inclusion material;
G_2—stiff modulus of rock;
v_1—Poisson's ratio of hollow inclusion material;
v_2—Poisson's ratio of rock;
ρ—the radial distance of the resistance foil gauge in the hollow inclusion;
$n = G_1/G_2$, $m = R_1/R_2$, $x_1 = 3-4v_1$, $x_2 = 3-4v_2$.

3.2 In-situ stress measurement and distribution law

3.2.1 The layout of in-situ stress measuring points

After selecting a proper method to measure the primary rock stress, it is of great engineering significance to correctly determine the number of measuring points and reasonably arrange the positions of measuring points, so as to ensure that the measured stress can accurately reflect the in-situ stress distribution law of the whole mining area from space. Therefore, each measuring point shall be carefully selected and the following principles shall be followed:

(1) The quality of rock mass. In order to ensure the integrity of the core

and the reliability of the in-situ stress measuring results, the surrounding rock mass around the measuring points shall be homogeneous and complete, and the drilling is positioned in this type of rock mass around the measuring points.

(2) Close to the object of study. For the mining area, the ore body is usually the object of study, so the measuring points shall be arranged in the ore body or its surrounding area. The measuring points shall be as close to the designed tunnel as possible. According to the geological structure data of the mining area, the measuring points shall be representative of the in-situ stress field of the designed tunnel.

(3) Avoid the nearby tunnel cavern engineering under construction, avoid the stress distortion zone, unstable area and interference source, in order to ensure the authenticity of the primary rock stress;

(4) Avoid the influence of fault exerted on the measured value. The measured results show that near large faults, not only the horizontal stress value is low, but also the direction of principal stress may be disturbed. Therefore, the measuring points shall be arranged as far away from the fault and crushed zone as possible.

(5) In order to study the law of in-situ stress status changing with depth, the measurement shall be carried out in three or more middle sections as far as possible.

In addition to the above principles, the actual conditions of the site shall also be considered, such as the handling and installation of the drill, ensuring the safety of water, electricity and ventilation, and not affecting the normal production.

According to these principles, on the basis of engineering geological investigation, the research group of in-situ stress measurement from the University of Science and Technology Beijing has combined with the site construction conditions of Sanshandao Gold Mine, selected the measuring points basically avoiding the bending, forking or turning, etc. areas of stress concentration of the tunnel and stope, as well as fault, rock crushed zone, and fracture development zone. At the same time, the measuring points are as far as possible away from large goaf and cavern. In addition, the measured drilling depth is more than 3 times of the tunnel span, and the measuring points are more than 50 m away from the adjacent tunnel or other excavation works, which ensure that all the measuring points are located in the primary rock stress area. Refer to Table 3-1 for the layout of each measuring point and description of the drilling condition.

The position of each measuring point of in-situ stress and the general condition of drilling construction　　　Table 3-1

No. of measuring point	Position	Coordinate(x,y,z)	Buried depth (m)	Hole depth (m)	RQD (%)
1	−510 m N tunnel	(41837.8,96016.9,−512.0)	512	8.10	48.5
2	−510 m S tunnel	(41032.3,95701.3,−512.5)	512.5	9.44	64.2
3	−555 m S tunnel	(40825.7,95716.6,−553.0)	553	8.63	81.2
4	−600 m near the internal shaft	(40804.2,95767.1,−602.2)	602.2	9.71	61.3
5	−600 m newly-built connection tunnel	(40760.6,95485.1,−603.0)	603	9.64	88.4
6	−645 m entrance of slope ramp	(40901.7,95813.3,−647.0)	647	9.13	47.4
7	−690 m entrance of slope ramp	(40979.5,95851.2,−693.0)	693	10.79	75.3
8	−690 m near the air shaft	(40825.6,95878.1,−693.0)	693	9.20	82.7
9	−750 m entrance of slope ramp	(40941.2,95867.8,−750.0)	750	8.51	66.5
10	−795 m transportation tunnel	(40877.8,95957.6,−795.0)	795	8.20	91.7
11	−825 m transportation tunnel	(40719.8,96020.5,−825.0)	825	12.25	75.9
12	−960 m transportation tunnel	(40796.8,95901.1,−960.0)	960	8.85	91.3

3.2.2　The steps for in-situ stress measurement

The instruments and equipment for in-situ stress measurement are the cores of the field test. Almost all of the test instruments and equipment are provided by the University of Science and Technology Beijing, except the field drill and drill pipe. Refer to Table 3-2 for the main instruments and equipment.

The main instruments and equipment　　　Table 3-2

Name	Model and main technical parameters	Specification	Qty.
Drill	MK all-hydraulic drill, with drill depth of 150 m	piece	1
Drill pipe	$\phi 50$	set	1
Large coring drill bit	$\phi 130 (\phi 127)$	pc	12
Small coring drill bit	$\phi 36$	pc	12
Grind drill bit	$\phi 130$	pc	2
Large rock core barrel	Matching with large drill bit	pc	6
Small rock core barrel	Matching with small drill bit	pc	2

continued

Name	Model and main technical parameters	Specification	Qty.
Digital dynamic signal testing system (collector)	Model DDS163, with measuring range of $+1000$ $\mu\varepsilon$; strain coefficient $K=2$; resolution ratio of 0. 1$\mu\varepsilon$; signal-to-noise ratio$\geqslant$52	piece	1
Inclusion type stress gauge	Model KX-0436100H, resistance value of foil gauge $123+0.2\Omega$, coefficient of sensitivity $2.06+1\%$	pc	12
Mounted rod for stress gauge		set	1
Calibrator for calibrating Calibrator for calibrating confining pressure	Model CW-250	piece	1
High-low temperature test chamber	Model SD-202, with accuracy of 0. 1℃	piece	1

According to the basic principle of stress relief method of hollow inclusion cell, the specific measurement test steps (Figure 3-4) are as follows:

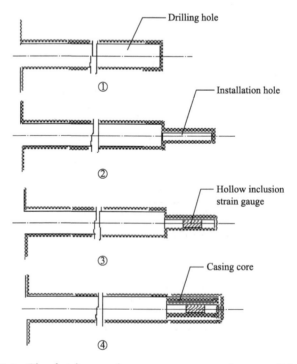

Figure 3-4 The sketch map of measurement steps of stress relief method

3 In-situ Stress Measurement Method and Its Application

(1) A horizontal drilling with a diameter of 130 mm is drilled on the tunnel wall at the selected measuring point till the depth with 3-5 times of the tunnel span. The specific relief position is determined by the situation of the drilling rock core, generally to the hole depth of 3-4 times of the tunnel span. The drilling is inclined updip 1°-3° to allow the cooling water to flow out and the drilling to be cleaned easily.

(2) Grind the bottom of the hole with a flat drill bit, hit the bell mouth with a conical drill bit, and then drill a concentric small hole with a diameter of 36 mm from the bottom of the hole, with the small hole depth of 35-40 cm. The quality of the bell mouth must be ensured because the bell mouth will play an important role in ensuring the concentricity of the small hole and the next step to make the hollow inclusion cell enter the small hole smoothly. After the small hole is washed with water, and the small hole is scrubbed back and forth by the inside wiping unit soaked with acetone, in order to completely remove the oil and other dirt in the small hole.

(3) The cemented agent (epoxy resin) and curing agent are mixed in proportion, stirred evenly, and injected into the cavity of the strain gauge. The plunger is fixed with a bolt, and then the plunger is sent into the predetermined position in the small hole with the mounted rod with an orientator. After the conical head at the front of the strain gauge hits the bottom of the small hole, the mounted rod is forcibly pushed to cut off the bolt, so that the plunger enters the cavity. The cementing agent in the cavity flows into the annular gap between the strain gauge and the drilling wall through the center hole of the plunger and the radial hole at the back. The sealing rings at both ends of the strain gauge will prevent the cementing agent to flow from this gap. When the cementing agent solidifies, the strain gauge is tightly cemented to the small hole.

(4) After the solidification of the cementing agent (usually about 20 hours), the stress relief test can be carried out. Before the stress relief, the wire cable of the strain gauge first passes through the rock core barrel, the drill pipe and its rear water line tee, and connects to the electric bridge conversion device. Then, a thin-walled drill bit with a diameter of 130 mm is used to continue to extend the large hole, so that the rocks around the strain gauge gradually separate from the surrounding rocks, so as to realize the stress relief of the trepan boring rock core. In the process of stress relief, the strain values measured by each foil gauge in the strain gauge are automatically recorded by the electric

bridge conversion device and the data collector. According to the instructions, strain data is recorded once every 2 cm footage.

After the stress relief is completed, the strain data stored in the data collector is printed out by the computer, and the stress relief curve is plotted accordingly, that is, the strain value of each foil gauge changes with the stress relief depth.

3.3 The in-situ stress field distribution

The Figure 3-5 shows the site stress relief curve at the horizontal measuring point of -795 m for in-situ stress measurement. It can be seen that in the process of overcoring relief, the variation of restoring strain is basically synchronous with footage depth. Before the relief depth of the trepan boring reaches the measured section (that is, the section where foil gauge is located), the strain value measured by each foil gauge is generally small. Some foil gauges even measure the negative strain values. This is the result of stress transfer caused by trepan boring, which is equivalent to "excavation effect". When the relief depth of trepan boring is close to the measured section, many curves change in the opposite direction. The maximum strain value occurs when the trepan boring drill bit passes near the measured section. When the depth of the trepan boring exceeds a certain distance of the measured section, the strain value gradually stabilizes and the curve tends to be stable. The final stable value will be used as the raw data to calculate the in-situ stress.

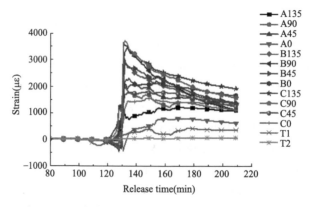

Figure 3-5   The stress relief curve at the horizontal measuring point of -795 m

At each measuring point, there are 12 foil gauges to measure strain values in 12 directions. A_{90}, A_0, A_{45}, A_{135}, B_{90}, B_0, B_{45}, B_{135}, C_{90}, C_0, C_{45}, C_{135} represent 12 foil gauges respectively, in which A, B and C represent three groups of strain rosettes. Each group is composed of four foil gauges. The subscript numbers (90, 0, 45, and 135) represent the included angles between the foil gauges and the directions of the drill axes.

In order to use the newly complete temperature compensation technology, the temperature calibration test of the trepan boring rock core must be carried out. The trepan boring rock core is placed in a constant temperature box with adjustable temperature. Connect the strain gauge wire into the electric bridge conversion device, adjust the initial reading to zero one by one, then raise the temperature by 5℃ every 4 hours, and record the strain value measured by each foil gauge in each temperature range. The data collection and recording are still completed automatically by the data collector. Because the thermistor is still located in the foil gauge of strain gauge in the trepan boring rock core, it is the most reliable and accurate way to measure the temperature of the foil gauge. The temperature in the constant temperature box is also monitored by a highly sensitive platinum film digital thermometer.

It takes approximately 4 hours for the temperature in thetrepan boring rock core uniformly, that is, the temperature at the foil gauge to be the same as the temperature in the constant temperature box. According to the recorded measured strain data and temperature value of each foil gauge when the temperature rises, the temperature strain rate of each foil gauge can be obtained, that is, the additional temperature strain value caused by the foil gauge is obtained when temperature rises by 1 ℃. The temperature change value of foil gauge can be obtained by converting the change of thermistor in hollow inclusion cell measured during stress relief. By multiplying the temperature strain rate of the temperature calibration test by the temperature change value of the foil gauge during the stress relief process, the additional strain value of each foil gauge due to the temperature change of the measuring point during the stress relief process can be obtained. By removing this additional strain value from the final stable strain value measured during the stress relief process, the authentic strain value caused by stress relief can be obtained (Figure 3-6).

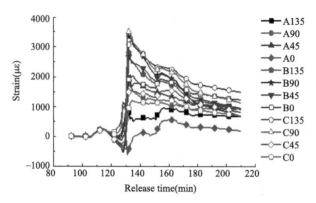

Figure 3-6 The relief curve with dual-temperature compensation and revision

The final stable strain value measured by each foil gauge at each measuring point Table 3-3

No. of measuring point	Strain value($\mu\varepsilon$)											
	A_0	A_{90}	A_{45}	A_{135}	B_0	B_{90}	B_{45}	B_{135}	C_0	C_{90}	C_{45}	C_{135}
1	280	587	407	343	233	858	58	962	263	771	919	77
2	426	598	342	465	426	986	156	1010	397	860	908	166
3	243	438	394	258	232	662	703	235	226	450	63	652
4	442	603	80	850	513	841	735	375	379	658	555	397
5	206	1399	1036	498	202	1025	308	984	197	551	402	357
6	319	1006	1214	—	319	759	240	767	303	694	91	892
7	240	516	635	129	255	1047	−186	1460	251	763	984	−96
8	200	846	929	142	189	860	226	764	195	783	380	631
9	345	1055	1099	240	282	1017	701	660	300	1021	182	1117
10	358	1132	1012	260	272	1172	1245	1183	1021	1377	1564	1089
11	512	1381	1064	1287	1546	978	1226	1256	1013	1482	1807	1139
12	216.1	1553.3	840.6	856.3	281.2	1553.0	734.4	1109.5	278.3	1573.4	918.3	991.0

The additional temperature strain rate of the foil gauge at each direction and each measuring point Table 3-4

No. of measuring point	The temperature strain rate of foil gauge at each channel($\mu\varepsilon/{}^\circ\!C$)											
	A_0	A_{90}	A_{45}	A_{135}	B_0	B_{90}	B_{45}	B_{135}	C_0	C_{90}	C_{45}	C_{135}
1	64	40	13	15	21	32	13	40	60	35	12	40
2	42	24	5	19	43	26	4	24	39	10	2	14

3 In-situ Stress Measurement Method and Its Application

continued

No. of measuring point	The temperature strain rate of foil gauge at each channel($\mu\varepsilon$/℃)											
	A_0	A_{90}	A_{45}	A_{135}	B_0	B_{90}	B_{45}	B_{135}	C_0	C_{90}	C_{45}	C_{135}
3	20	52	5	37	27	39	24	32	22	33	315	13
4	35	25	5	32	50	31	10	24	20	30	5	30
5	17	20	15	17	18	17	29	33	16	15	18	21
6	34	20	4	16	35	21	3	20	32	8	2	11
7	31	-27	1	5	50	-42	-9	12	37	32	-8	0
8	22	45	17	38	16	31	29	0	18	30	21	23
9	37	73	56	54	17	53	40	36	22	59	36	34
10	55	50	53	45	30	43	58	39	27	21	22	96
11	47	58	56	52	27	15	33	62	36	-4	27	63
12	97.4	141.1	117.7	118.9	90.5	136.5	108.4	112.8	91.1	141.9	120.0	118.1

It can be seen from the results of temperature calibration test (Table 3-3, Table 3-4) that the strain values of most foil gauges increase with the rises of temperature, and the temperature strain rates are different at different temperature ranges. The temperature strain rates consistent with the temperature ranges used in the underground test must be used in the calculation. The records of temperature changes at each measuring point in the process of stress relief are usually 1-2 ℃. This temperature change value multiplies by the temperature strain rate of foil gauge at each measuring point that concludes the additional temperature strain values (absolute values) of many foil gauges caused by temperature changes at the measuring points during the stress relief process, which can reach 50-100 microstrain, some even more than 100 microstrain. The measured total strain values in the stress relief process is almost in the same quantity magnitude. It shows that temperature is an important factor for in-situ stress measurement. The correct method to compensate or revise the influence of temperature is of great significance to ensure the accuracy of in-situ stress measuring results. After the stable strain values measured in the stress relief process are revised by the temperature influence, the strain values of the hollow inclusion cell in all directions due to the stress relief can be obtained, which are the final strain values used to calculate the in-situ stress (Table 3-5).

The final strain values of each measuring point for calculating in-situ stress Table 3-5

No. of measuring point	Strain value($\mu\varepsilon$)											
	A_0	A_{90}	A_{45}	A_{135}	B_0	B_{90}	B_{45}	B_{135}	C_0	C_{90}	C_{45}	C_{135}
1	190	531	389	322	203	813	40	905	178	722	902	21
2	227	485	318	375	222	863	137	897	213	813	899	100
3	242	435	393	256	231	660	702	233	225	448	45	652
4	300	501	60	720	310	715	694	278	298	536	534	275
5	146	1329	984	438	139	966	206	869	141	499	339	283
6	251	966	1207	—	249	718	234	728	240	678	87	870
7	195	555	634	122	184	1108	−173	1443	198	717	995	−96
8	195	835	926	133	185	853	219	764	191	776	375	626
9	250	867	955	101	239	880	597	568	243	869	89	1029
10	260	489	842	378.5	256	856	625	312	260	1264	987	1121
11	364	876	1202	1121	1464	1184	876	1055	901	—	—	935
12	131	1430	738	752	202	1434	640	1011	199	1449	814	888

It is necessary to knowthe rock elastic modulus and Poisson's ratio (μ) at the measuring point when calculating the in-situ stress from the drilling strain value measured by the stress relief test. It is the most scientific and economical method to determine the rock elastic modulus and Poisson's ratio through the confining pressure calibration test of trepan boring rock core. Because after completion of the stress relief, the hollow inclusion cell is still cemented within the center of small hole. By the application of confining pressure on the trepan boring rock core, the strain values measured by hollow inclusion cell caused by confining pressure, according to the measured confining pressure—strain curve, based on the theory of thick wall cylinder, that can be calculated the rock elastic modulus and Poisson's ratio by the equation. The calculation of in-situ stress requires the rock elastic modulus and Poisson's ratio at each measuring point, but the state and mechanical properties of underground rock mass are extremely complex and ever-changing. Their properties may vary greatly from one point to another not far away from each other. In order to ensure the accuracy of the in-situ stress calculation results, the used quantity of the rock elastic modulus and Poisson's ratio must be truly representative of foil gauge rock found at that site.

Since the confining pressure calibration test is carried out by using the primary rock core with relief from the site trepan boring, the results calculated from the confining pressure test can fully ensure this requirement, which is of great significance for accurate measurement and calculation of in-situ stress.

The testis carried out on a calibrator for calibrating confining pressure, which consists of an oil hydraulic pump and a cylinder oil hydraulic cylinder (Figure 3-7). The steps of confining pressure calibration test are as follows:

(1) Place the trepan boring rock core in the cylinder oil hydraulic cylinder of the calibrator for calibrating confining pressure, and make the foil gauge locate in the middle of the cylinder; Connect the wire of the strain gauge to the electric bridge conversion device, and adjust the initial strain reading of each channel to zero one by one;

Figure 3-7 The high-pressure calibrator for calibrating confining pressure

(2) The oil hydraulic pump is used manually to pressurize the trepan boring rock core, and the strain values of each foil gauge in the hollow inclusion cell caused by confining pressure are automatically collected and recorded by the data collector according to the instructions. Readings are recorded for every 2 MPa increase in pressure until the pressure increases to 12-16 MPa. Then implement pressure relieves, the readings are also recorded for every 2 MPa in pressure relief;

(3) After the complete pressure relief, the pressure increase-pressure relief process mentioned above is repeated once, and the pressure relief trend of this process is generally used to calculate the elastic modulus and Poisson's ratio.

If the rock is continuous, homogeneous and isotropic, andeach foil gauge of the strain gauge works normally and reliably, the strain values in the same angle direction with the drilling axis shall be very close. The circumferential strain is compression strain, the axial strain is tensile strain, and the slope strain is compression strain, but the strain value is less than the circumferential strain.

Therock elastic modulus, Poisson's ratio and three dimensional in-situ stress value at each measuring point are obtained from the stress relief strain values calibrated by temperature and the confining pressure calibration test results. The calculated results are shown in Table 3-6 and Table 3-7 respectively.

The elastic modulus (E), Poisson's ratio (μ) and K coefficient values at each measuring point   Table 3-6

No. of measuring point	E(GPa)	μ	K_1	K_2	K_3	K_4
1	50	0.25	1.045	1.061	1.030	0.978
2	52	0.22	1.120	1.138	1.076	0.917
3	56	0.28	1.139	1.158	1.089	0.956
4	60	0.21	1.141	1.157	1.088	0.891
5	55	0.25	1.155	1.166	1.098	0.924
6	55	0.20	1.122	1.140	1.077	0.894
7	50	0.36	1.132	1.556	1.087	1.000
8	56	0.28	1.139	1.158	1.089	0.956
9	51	0.35	1.148	1.159	1.091	0.824
10	45	0.25	1.019	1.000	1.012	0.991
11	39.33	0.26	1.199	1.162	1.127	0.913
12	50.83	0.20	1.137	0.997	1.089	0.935

The calculated results of principal stress at different depths   Table 3-7

Depth (m)	Maximum principal stress σ_1			Intermediate principal stress σ_2			Minimum principal stress σ_3		
	Value (MPa)	Direction (°)	Dip angle (°)	Value (MPa)	Direction (°)	Dip angle (°)	Value (MPa)	Direction (°)	Dip angle (°)
75	6.01	288.5	−6.3	3.81	198	−4.9	2.56	250.4	82
150	7.73	280.9	−5.2	5.48	9.4	16.6	4.5	27.7	72.5
420	19.27	284.1	−21.3	11.05	18.5	−11.1	10.88	134.4	−65.7
420	19.39	120.4	−14.9	10.92	169.2	68.1	9.44	34.7	15.8
510	24.55	129	4	14.49	133	−85	16.35	−138	2
510	24.64	−111	3	15.68	155	82	15.02	161	−10
555	25.71	−45	−13	14.00	14	73	13.00	50	−20
600	28.88	103	1	16.54	10	76	14.77	13	−8
600	30.17	110	−16	16.94	236	−70	18.83	24	−11
645	29.57	112	−3	19.56	−177	−80	15.48	−156	−9
690	31.50	−80	2	19.08	230	−79	17.54	10	−10
690	29.77	−83	4	20.84	−8	−74	19.63	8	15

continued

Depth (m)	Maximum principal stress σ_1			Intermediate principal stress σ_2			Minimum principal stress σ_3		
	Value (MPa)	Direction (°)	Dip angle (°)	Value (MPa)	Direction (°)	Dip angle (°)	Value (MPa)	Direction (°)	Dip angle (°)
750	33.22	119	−10	19.93	−89	−82	17.10	208	−8
795	30.72	133.3	−14.9	18.09	149.7	74.5	26.41	−135.6	−4.17
825	48.93	164.09	3	23.15	74.41	−5.97	21.66	47.22	83.29
960	45.31	60.33	7.07	43.79	−30.80	9.10	33.48	7.66	−78.44

In order to more accurately reflect the variation law of the primary rock stress field with the direction of depth in the Sanshandao Gold Mine, this study combines with the measuring results of all the in-situ stress in the stage (Table 3-7), a total of 12 measuring points are analyzed and studied, and linear regression is carried out for them. The regression equations and regression curves of the maximum horizontal principal stress, the minimum horizontal principal stress and the vertical principal stress with depth are obtained (Figure 3-8).

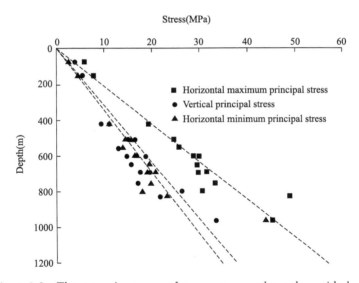

Figure 3-8 The regression curves of $\sigma_{h,\max}$, $\sigma_{h,\min}$ and σ_v values with depth

Through the regression analysis of the measured horizontal in-situ stress data, the shallow in-situ stress field model at −960 m in Xishan Mining Area of Sanshandao Gold Mine is:

Regression equations:

$$\sigma_{h,\max}=0.048H \tag{3-6}$$

$$\sigma_{h,min} = 0.029H \qquad (3\text{-}7)$$
$$\sigma_v = 0.032H \qquad (3\text{-}8)$$

where: $\sigma_{h,max}$, $\sigma_{h,min}$, and σ_v are respectively the maximum horizontal principal stress, the minimum horizontal principal stress and the vertical principal stress, MPa; H is the depth, m.

4

Rock Mass Investigation and Rock Quality Evaluation

4.1 Influence of structural plane on surrounding rocks

With the existence of structural plane, the characteristics of engineering rock mass are changed, the intactness of rock mass is damaged, the strength is reduced, the stress distribution is uneven, and the anisotropy is enhanced. The practice shows that the instability of engineering rock mass is closely related to the development degree and combination features of structural plane. The stress state of rock mass is the precondition of rock mass failure, but its action shall be manifested through the mechanical effect of rock mass structure. The properties and combination forms of structural planes often determine the deformation and failure modes of rock mass. Due to the existences of structural planes, the deformation, movement and failure of the surrounding rocks usually occur along the structural planes when excavating tunnel or mining in rock mass, as shown in Figure 4-1.

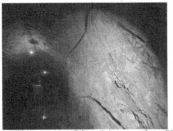

(a) The middle section of the mining tunnel at −780 m   (b) The middle section of the mining tunnel at −780 m

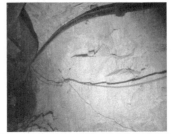

(c) The middle section of the slope tunnel at −825 m   (d) The middle section of the slope tunnel at −870 m

Figure 4-1 The wall caving phenomenon of tunnel surrounding rocks

In the complex geological environment of "three-high and one-disturbance", the influence of dip angle, density of structural plane, combination relationship and direction of principal stress on the deformation and failure law of tunnel surrounding rock mass is more important than the property of rock itself. It is a fundamental and preconditioning work to study the structural plane features of tunnel surrounding rock mass, which provides basic data support for the studies of tunnel rock mass mechanical characteristics, tunnel stability and tunnel support.

4.2 Structural plane investigation of deep tunnel rock mass

ShapeMetriX 3D is a set of 3G software and measurement products manufactured by the Austria 3 GSM Company, which is a new, and the highest level 3D contactless measurement system with geometrical parameters of rock mass. It applies to geotechnical engineering, engineering geology and survey, and builds rock mass and true 3D digital model of topographical surface. Also, it provides the corresponding software analysis system of 3D digital model for processing, in order to get a full and detailed geometry measurement data of rock mass. This measurement system records the side slope, tunnel contour and spacing position and occurrence of actual rock mass discontinuity surface, determines the spacing geometry mining stope, and determines the amount of excavation, identification of dangerous rock mass stability, and block movement analysis, etc.

This system is suitable for scientific research, widely used in computer analysis laboratories, simulation centers, and engineering computing centers, etc. It is a new system that can obtain 3D image surfaces and completely preserve all the site data for future indoor measurement and analysis. It is generally used in tunnel excavation, mining, different types of documentations, truly recording and saving the site 3D images and data, and for 3D processing, measurement, and analysis (Figure 4-2).

Taking -780 m horizontal measuring point 1 as an example, the information collection process of rock mass structural plane (Figure 4-3) is as follows:

(1) The position of measuring point

The measuring point is located at the entrance of the middle section of the

4　Rock Mass Investigation and Rock Quality Evaluation

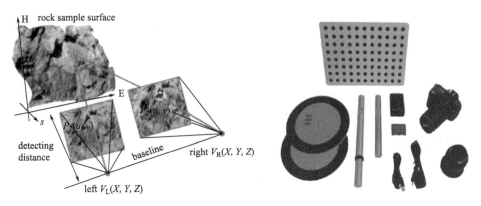

(a) The principle of 3D rock mass surface measurement　　(b) The 3D contactless measurement system

Figure 4-2　The principle of digital recognition of structural plane of joint rock mass and the main instrument sketch

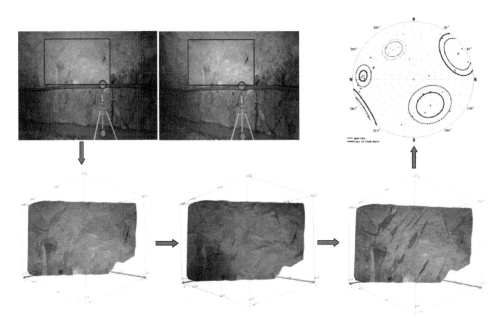

Figure 4-3　The information collection process of rock mass structural plane

mining tunnel at -780 m. The surrounding rock of the tunnel is not supported by gunite, and the structural plane is obviously developed. Coordinates (X, Y, Z) are $(40967.89, 95943.34, -780)$, the air temperature around the measuring point is 34 ℃, and the humidity is 83%.

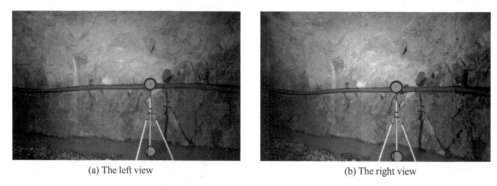

(a) The left view　　　　　　　　(b) The right view

Figure 4-4　The image of rock mass surface

(2) The feature analysis ofstructural plane

The left view and right view from the site are as shown in Figure 4-4. Import both the left and right views into the ShapeMetrix 3D software analysis system, marking out the key measurement area. The system synthesizes 3D model and makes azimuth and distance real by a series of technologies such as pixel matching and anamorphose deviation revision system, and gets the 3D view of rock mass surface, as shown in Figure 4-5.

On the synthesized 3D view, the rock mass joints are grouped according to the distribution of major joints and fissures. Different colors represent different groups. The main fissures are divided into four groups of red, green, blue and yellow, as shown in Figure 4-6.

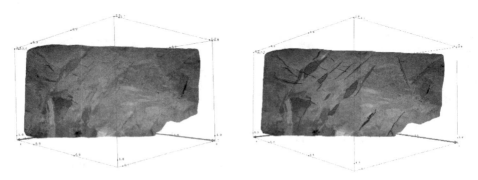

Figure 4-5　The synthesized 3D view　　Figure 4-6　The distribution of joints in the model

As shown in Figure 4-7, the system draws the equivalent density sketch of measuring points and poles according to the spacing distribution and grouping of structural planes. According to the pole equivalent density sketch, the trend and dip angle distribution information of all the structural planes can be determined.

The advantages of occurrence are 94.41°∠72.08°, 328.00°∠44.17°, 239.23°∠75.89°, 148.11°∠51.92°.

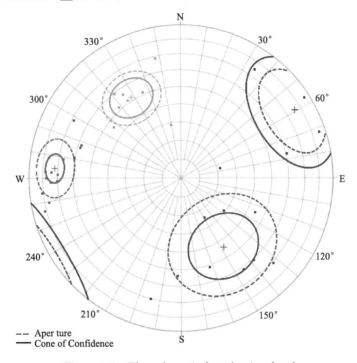

Figure 4-7 The pole equivalent density sketch

According to the spacing distribution and grouping of each group of structural plane, the system draws the structural plane interval sketch as shown in Figure 4-8.

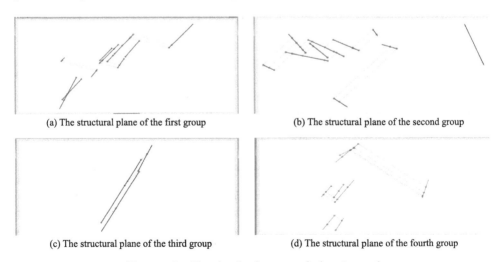

Figure 4-8 The sketch of structural plane interval

On the basis of the synthesized 3D view, the joint area and path line of the structural plane are calibrated according to the synthesized 3D rock surface. The system automatically analyzes the trend, dip angle, path line length and interval of the structural planes, and automatically records the parameters of the structural planes with the data, as shown in Table 4-1.

The statistics of each joint set parameter   Table 4-1

No. of joint set	Qty. of joint (line)	Frequency of joint line(m)	Average interval (m)	Maximum interval (m)	Total length of joint (m)	Average length (m)
First	10	3.69	0.27	0.73	3.85	0.385
Second	10	4.32	0.23	0.83	2.63	0.263
Third	3	4.21	0.24	0.31	1.21	0.403
Fourth	8	2.24	0.45	1.35	2.68	0.335

According to the results of the software post-processing, the geometrical information of the obtained joints is grouped and counted in the Table 4-2 to Table 4-5.

The geometrical parameters of the first joint set   Table 4-2

No. of path line	Length of path(m)	Trend(°)	Dip angle(°)
1	0.393	99.5	68.0
2	0.463	101.1	61.3
3	0.650	107.1	59.4
4	0.778	118.4	69.5
5	0.653	108.4	59.0
6	0.563	77.0	81.6
7	0.082	92.1	70.8
8	0.221	99.1	77.5
9	0.294	89.1	71.0
10	0.174	84.4	81.8

The geometrical parameters of the second joint set   Table 4-3

No. of path line	Length of path(m)	Trend(°)	Dip angle(°)
1	0.196	342.3	50.6
2	0.297	2.7	52.9

continued

No. of path line	Length of path(m)	Trend(°)	Dip angle(°)
3	0.289	294.5	43.6
4	0.445	306.8	28.1
5	0.146	15.0	69.5
6	0.516	254.1	21.3
7	0.096	318.2	60.0
8	0.166	329.2	23.8
9	0.604	312.3	70.1
10	0.329	325.6	60.5

The geometrical parameters of the third joint set Table 4-4

No. of path line	Length of path(m)	Trend(°)	Dip angle(°)
1	0.275	256.5	60.2
2	0.496	236.4	86.0
3	0.519	213.9	74.9

The geometrical parameters of the fourth joint set Table 4-5

No. of path line	Length of path(m)	Trend(°)	Dip angle(°)
1	0.510	127.2	46.9
2	0.281	168.3	29.6
3	0.348	155.5	52.2
4	0.536	145.2	52.6
5	0.304	144.9	56.9
6	0.273	138.6	52.5
7	0.274	141.3	60.3
8	0.389	167.6	69.7

The density of structural plane: $\lambda_1 = 2.000$ line/m^2; $\lambda_2 = 2.000$ line/m^2; $\lambda_3 = 0.600$ line/m^2; $\lambda_4 = 1.600$ line/m^2.

4.3 Investigation summary of the structural plane of deep tunnel rock mass

The joint features obtained from the investigation are summarized in Table 4-6 according to their locations.

The summary of surrounding rock investigation of the deep tunnel   Table 4-6

No.	Coordinate	Prospecting area	3D view of surrounding rock	Occurrence of structural plane
1	X=40967.89 Y=95943.34 Z=-780			94.41°∠72.08° 148.11°∠51.92° 239.23°∠75.89° 328.00°∠44.17°
2	X=40996.32 Y=95952.43 Z=-780			88.00°∠54.81°
3	X=40895.37 Y=95955.89 Z=-795			101.68°∠68.82° 342.74°∠53.31°
4	X=40875.93 Y=95950.98 Z=-795			29.02°∠78.63° 148.37°∠55.23° 316.06°∠64.43°
5	X=40837.19 Y=96003.17 Z=-825			259.14°∠79.53° 328.33°∠64.50°
6	X=40850.77 Y=95997.05 Z=-825			66.40°∠72.38° 313.36°∠60.85°

■ 4 Rock Mass Investigation and Rock Quality Evaluation ■

continued

No.	Coordinate	Prospecting area	3D view of surrounding rock	Occurrence of structural plane
7	$X=40850.77$ $Y=95997.05$ $Z=-825$			15.31°∠56.50° 88.02°∠60.48° 202.35°∠80.53°
8	$X=40845.17$ $Y=96006.77$ $Z=-825$			90.28°∠61.68° 191.22°∠85.07° 318.50°∠72.55°
9	$X=40900.00$ $Y=96044.65$ $Z=-870$			41.40°∠81.97° 107.45°∠71.64°
10	$X=40773.00$ $Y=96016.02$ $Z=-885$			234.24°∠51.04° 327.47°∠81.97°
11	$Z=-900$			194.40°∠66.62° 166.15°∠57.77°
12	$X=40796$ $Y=96078$ $Z=-915$			144.61°∠68.25° 279.97°∠55.92°

continued

No.	Coordinate	Prospecting area	3D view of surrounding rock	Occurrence of structural plane
13	X=40796 Y=96079 Z=−915			140.52°∠69.98° 312.73°∠55.56°

The distribution and occurrence of each deep horizontal joints and fissures and the statistical analysis results are shown in Table 4-7 to Table 4-9.

The list of deep joints and fissures   Table 4-7

Depth (m)	Position	Qty. of joint set	Qty. of joint	Average interval (cm)	Surface density line(m²)	Advantage of occurrence of structural plane
−780	S17162 Entrance of mining	1	52	25	5.14	88°∠54.81°
	The middle section of horizontal tunnel	1	10	27	2	94°∠72°
		2	10	23	2	328°∠44°
		3	3	24	0.6	239°∠76°
		4	8	45	1.6	148°∠52°
−795	20 m S at the entrance of slope tunnel	1	18	20	1.64	102°∠69°
		2	14	21	1.27	343°∠53°
	40 m S at the entrance of slope tunnel	1	19	48	1.98	148°∠55°
		2	9	23	0.94	29°∠79°
		3	5	39	0.52	316°∠64°
−825	40 m S at the entrance of slope tunnel	1	6	63	0.88	328°∠65°
		2	3	58	0.44	259°∠80°
	35 m N at the entrance of slope tunnel	1	10	15	1.53	66°∠72°
		2	19	18	2.53	313°∠61°
	35 m N at the entrance of slope tunnel	1	2	—	0.33	202°∠81°
		2	2	—	0.33	15°∠57°
		3	13	23	2.17	88°∠60°
	44 m S at the entrance of slope tunnel	1	5	50	1.02	90°∠62°
		2	4	26	0.82	319°∠73°
		3	6	33	1.23	191°∠85°

4 Rock Mass Investigation and Rock Quality Evaluation

continued

Depth (m)	Position	Qty. of joint set	Qty. of joint	Average interval (cm)	Surface density line(m²)	Advantage of occurrence of structural plane
−870	At the entrance of slope tunnel	1	7	111	1.28	107°∠72°
		2	7	83	1.28	41°∠82°
−885	Slope tunnel	1	26	48	1.21	234°∠51°
		2	17	29	0.92	327°∠54°
−900	Slope tunnel	1	21	32	1.11	208°∠69°
		2	14	16.4	3.21	161°∠60°
−915	7 m N at the entrance of slope tunnel	1	13	65	1.33	141°∠70°
		2	20	44	—	282°∠52°
	10 m S at the entrance of slope tunnel	1	17	65	0.68	141°∠76°
		2	15	49.9	0.96	313°∠56°

The statistical analysis of deep joint trend of each section   Table 4-8

Grouping of trend	−780 m middle section		−825 m middle section		−870 m middle section		−915 m middle section	
	Average dip angle(°)	Percentage (%)	Average dip angle(°)	Percentage (%)	Average dip angle(°)	Percentage (%)	Average dip angle(°)	Percentage (%)
0-10	52.9	1.2	77.6	1.4	—	0	—	0
10-20	69.5	1.2	8	1.4	—	0	—	0
20-30	—	0	—	0	52.3	14.3	—	0
30-40	—	0	38.4	1.4	—	0	—	0
40-50	35.5	1.2	59.5	1.4	83.9	7.1	—	0
50-60	45.2	2.4	79.1	2.9	83.6	7.1	—	0
60-70	46.5	4.9	77.7	5.7	—	0	—	0
70-80	60.1	8.5	63.5	7.1	55.1	7.1	—	0
80-90	53.2	19.5	68.7	8.6	—	0	—	0
90-100	62.2	29.2	59.5	11.4	83.5	7.1	—	0
100-110	64.5	8.5	47.9	2.9	62.3	21.4	—	0
110-120	63.8	2.4	—	0	—	0	70.6	2.1
120-130	46.9	1.2	—	0	—	0	56.4	2.8
130-140	52.5	1.2	—	0	71.8	7.1	58.5	4.9
140-150	56.6	3.7	—	0	—	0	54.4	7.7
150-160	52.2	1.2	—	0	—	0	59.8	7.0

■ Mining Environment and Rockmass Control in Deep Coastal Mine ■

continued

Grouping of trend	−780 m middle section		−825 m middle section		−870 m middle section		−915 m middle section	
	Average dip angle(°)	Percentage (%)	Average dip angle(°)	Percentage (%)	Average dip angle(°)	Percentage (%)	Average dip angle(°)	Percentage (%)
160-170	49.7	2.4	89.4	1.4	—	0	60.6	6.3
170-180	—	0	84	2.9	—	0	66.0	3.5
180-190	—	0	—	0	—	0	67.1	2.8
190-200	—	0	—	0	77.4	7.1	68.7	2.8
200-210	—	0	81.1	4.3	87.1	7.1	71	2.8
210-220	74.9	1.2	86.5	2.9	—	0	60.3	4.2
220-230	—	0	—	0	—	0	49.4	4.9
230-240	86	1.2	—	0	—	0	56.4	4.2
240-250	—	0	—	0	82	7.1	40.6	2.8
250-260	41.7	3.6	76.8	2.9	—	0	53.3	3.5
260-270	—	0	—	0	—	0	49.7	4.9
270-280	—	0	71.1	2.9	—	0	38.2	2.1
280-290	—	0	69.5	1.4	—	0	60.5	4.2
290-300	28.1	1.2	49.6	7.1	68.1	7.1	66.5	2.8
300-310	65.1	2.4	64.7	8.6	—	0	68.5	2.8
310-320	42.2	2.4	67.2	2.9	—	0	50.1	7.0
320-330	—	0	69.1	8.6	—	0	52.3	5.6
330-340	50.6	1.2	66.3	4.3	—	0	50.4	3.5
340-350	—	0	70.4	2.9	—	0	54.9	3.5
350-360	—	0	68	1.4	—	0	47.9	1.4

The statistical analysis of deep joint dip angle of each section   Table 4-9

Grouping of dip angle	−780 m	−795 m	−825 m	−870 m	−885 m	−900 m	−915 m
	Percentage(%)						
0-10	0	0	1.4	0.0	0	0	0
10-20	0	0	0	0.0	4.7	0	1.5
20-30	4.8	1.5	0	7.1	9.3	0	10.8
30-40	3.6	6.3	4.3	0	11.6	2.9	9.2
40-50	14.5	7.7	4.3	7.1	20.9	8.6	15.4
50-60	62.7	27.7	15.9	14.3	20.9	22.9	21.5
60-70	2.4	18.5	33.3	7.1	16.3	31.4	21.5

4 Rock Mass Investigation and Rock Quality Evaluation

continued

Grouping of dip angle	−780 m	−795 m	−825 m	−870 m	−885 m	−900 m	−915 m
	Percentage(%)						
70-80	0	23.1	17.4	14.3	9.3	20.0	13.8
80-90	1.2	13.8	23.2	50	7.0	14.3	6.2

The statistical analysis of the dip angle of deep joints and fissures is shown in Table 4-10 and Table 4-11.

The statistical analysis of deep joints and fissures trend Table 4-10

Trend(°)	Percentage(%)	Qty. of fissure(line)
0-10	0.55	2
10-20	1.37	5
20-30	1.37	5
30-40	0.82	3
40-50	1.09	4
50-60	1.64	6
60-70	2.19	8
70-80	3.28	12
80-90	6.28	23
90-100	9.84	35
100-110	5.74	21
110-120	1.91	7
120-130	2.73	10
130-140	3.83	14
140-150	3.55	13
150-160	4.92	18
160-170	3.28	12
170-180	2.73	10
180-190	1.09	4
190-200	1.64	6
200-210	1.91	7
210-220	3.01	11
220-230	2.46	9
230-240	2.19	8
240-250	1.37	5
250-260	1.91	7
260-270	1.91	7
270-280	1.09	4

continued

Trend(°)	Percentage(%)	Qty. of fissure(line)
280-290	2.19	8
290-300	2.73	10
300-310	3.28	11
310-320	3.83	14
320-330	4.92	18
330-340	3.55	13
340-350	3.06	11
350-360	2.46	9

The statistical analysis of dip angle of deep joints and fissures   Table 4-11

Dip angle(°)	Percentage(%)	Qty. of fissure(line)
0-10	0.27	1
10-20	1.12	3
20-30	4.87	13
30-40	7.36	11
40-50	13.90	50
50-60	23.98	87
60-70	23.98	89
70-80	14.44	53
80-90	13.08	48

The structural plane information of surrounding rock of tunnel is classified and statistically analyzed, and the trend zone is shown in Figure 4-9 and the dip angle zone in Figure 4-10.

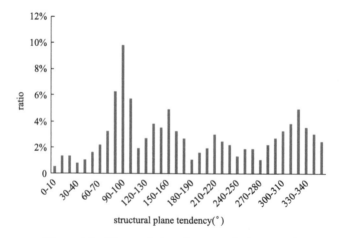

Figure 4-9   The statistics of structural plane trend

4 Rock Mass Investigation and Rock Quality Evaluation

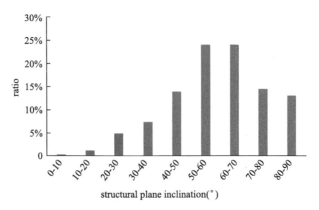

Figure 4-10 The statistics of dip angle of structural plane

The structural plane of tunnel surrounding rock with the trend of 70°-110° is about 26.05%, 130°-170° accounting for about 15.58%, 200°-240° accounting for about 9.57%, and 290°-350° accounting for about 18.09%. The dip angle mainly ranges from 40° to 80°, and accounting for about 76.3%.

4.4 Rock quality classification

4.4.1 Basic Quality classification

The BQ value of the Basic Quality Index of Rock Mass (also known as the BQ Classification Method of the National Standard) is put forward on the basis of summarizing the existing engineering rock mass classification methods of various industries in China. This classification takes the intactness of rock mass and the hardness of rock mass, which determine the basic characteristics of engineering rock mass quality, as the basic factors of classification. If the basic quality of rock mass is good, its stability is good. Otherwise, the stability is poor. The partitions of rock indexes are shown in Table 4-12 and Table 4-13.

The partition table of rock hardness   Table 4-12

Rock-saturated uniaxial compressive strength(MPa)	<5	5-15	15-30	30-60	>60
Hardness	Extreme soft rock	Soft rock	Relatively soft rock	Relatively hard rock	Hard rock

The partition of rock intactness degree   Table 4-13

Intactness index of rock mass K_v	<0.15	0.15-0.35	0.35-0.55	0.55-0.75	>0.75
Intactness degree	Extremely broken	Broken	Relatively broken	Relatively intact	Intact

According to the statistical investigation of joints and fissures in the previous period, the quantity of joints of rock mass volume J_v can be obtained, and the intactness index of rock mass K_v can be calculated according to J_v.

The intactness index of rock mass can be calculated based on the structural plane parameters of rock mass, and the calculation equations are as follows:

$$\left.\begin{array}{ll} K_v=1.0-0.083J_v & J_v \leqslant 3 \\ K_v=0.75-0.029(J_v-3) & 3 \leqslant J_v \leqslant 10 \\ K_v=0.55-0.02(J_v-10) & 10 \leqslant J_v \leqslant 20 \\ K_v=0.35-0.013(J_v-20) & 20 \leqslant J_v \leqslant 35 \\ K_v=0.15-0.0075(J_v-35) & J_v>35 \end{array}\right\} \quad (4-1)$$

where J_v—the quantity of joints of rock mass volume refers to the quantity of joints (structural planes) contained in an unit volume, which can be calculated by the following equations:

$$J_v = N_1/L_1 + N_2/L_2 + \cdots + N_n/L_n \text{ (line/m}^3) \quad (4-2)$$

where $L_1, L_2, \cdots, L_n$—the measured length perpendicular to the structural plane;

$N_1, N_2, \cdots, N_n$—the quantity of structural planes in the same group.

The *BQ* expression of the Basic Quality Index of Rock:

$$BQ = 90 + 3Rc + 250K_v \quad (4-3)$$

The uniaxial compressive strength of deep rock has been selected with reference to the test data of previous projects, and the basic quality indexes of rock calculated are shown in Table 4-14.

The basic quality indexes of rock mass   Table 4-14

No. of measuring point	Intactness index of rock mass(K_v)	Basic quality of rock(*BQ*)
−780 m horizontal measuring point 1	0.461	475.25
−780 m horizontal measuring point 2	0.720	540.00
−795 m horizontal measuring point 1	0.555	498.75
−795 m horizontal measuring point 2	0.575	503.75

■ 4 Rock Mass Investigation and Rock Quality Evaluation ■

continued

No. of measuring point	Intactness index of rock mass(K_v)	Basic quality of rock(BQ)
−825 m horizontal measuring point 1	0.741	545.25
−825 m horizontal measuring point 2	0.502	485.50
−825 m horizontal measuring point 3	0.713	538.25
−825 m horizontal measuring point 4	0.579	504.75
−870 m horizontal	0.693	533.25
−885 m horizontal	0.401	460.25
−900 m horizontal	0.412	463.00
−915 m horizontal measuring point 1	0.453	477.35
−915 m horizontal measuring point 2	0.481	484.35

According to the structural features and basic quality indexes of rock mass and with the reference to the classification standard in Table 4-15, the partition types of the stability of surrounding rock are shown in Table 4-16.

The classification standard of the basic quality of rock mass   Table 4-15

Classification of the basic quality	Qualitative feature of the basic quality of rock mass	Basic quality index of rock mass BQ
I	Extremely hard rock, with intact rock mass	>550
II	Extremely hard-hard rock, with relatively intact rock mass; Relatively hard rock, with intact rock mass	550-450
III	Extremely hard-hard rock, with relatively broken rock mass; Relatively hard rock or interbedded with the soft and hard rocks, with relatively intact rock mass; Relatively soft rock, with intact rock mass	450-350
IV	Extremely hard-hard rock, with broken rock mass; Relatively hard rock, with relatively broken-broken rock mass; Mainly the relatively soft rock or soft rock interbedded with the soft and hard rocks, with relatively intact-relatively broken rock mass; Soft rock, with intact-relatively intact rock mass	350-250
V	Relatively soft rock, with broken rock mass; Soft rock, with relatively broken or broken rock mass; All the extremely soft rock and all the extremely broken rock	<250

The classification of rock mass stability   Table 4-16

Position	Basic quality index of rock mass(BQ)	Qualitative grade
−780 m horizontal measuring point 1	475.25	II

continued

Position	Basic quality index of rock mass(BQ)	Qualitative grade
−780 m horizontal measuring point 2	540.00	II
−795 m horizontal measuring point 1	498.75	II
−795 m horizontal measuring point 2	503.75	II
−825 m horizontal measuring point 1	545.25	II
−825 m horizontal measuring point 2	485.50	II
−825 m horizontal measuring point 3	538.25	II
−825 m horizontal measuring point 4	504.75	II
−870 m horizontal	533.25	II
−885 m horizontal	460.25	II
−900 m horizontal	463.00	II
−915 m horizontal measuring point 1	477.35	II
−915 m horizontal measuring point 2	484.35	II

In addition, the basic quality index of rock mass can be revised by considering various influencing factors such as groundwater state (K_1), occurrence of structural plane (K_2) and natural in-situ stress (K_3) in combination with engineering characteristics, which can be used as quantitative basis for classification of different engineering bodies. The revision value $[BQ]$ of underground engineering is calculated according to the following equation:

$$[BQ] = BQ - 100(K_1 + K_2 + K_3) \tag{4-4}$$

where K_1—the revision coefficient influenced by groundwater;

K_2—the revision coefficient influenced by the occurrence of the main structural plane;

K_3—the revision coefficient influenced by the natural in-situ stress.

For each revision coefficient, please refer to Table 4-17 to Table 4-19.

The revision coefficient influenced by groundwater (K_1)   Table 4-17

State of groundwater \ BQ	>450	350-450	250-350	<250
Humid or dripping state	0	0.1	0.2-0.3	0.4-0.6
Water drenching or water inrush, hydraulic pressure ≤0.1 MPa or specific yield of 10 L/min	0.1	0.2-0.3	0.4-0.6	0.7-0.9

continued

State of groundwater \ BQ	>450	350-450	250-350	<250
Water drenching or water inrush, hydraulic pressure $\geqslant$ 0.1 MPa or specific yield of 10 L/min	0.2	0.4-0.6	0.7-0.9	1

The revision coefficient influenced by the occurrence of the main weak structural plane (K_2)   Table 4-18

The occurrence of structural plane and its combination relationship with cavern axis	The strike of structural plane and the included angle with tunnel axis $\alpha \leqslant 30°$, with dip angle of $\beta=30\text{-}75°$	The strike of structural plane and the included angle with tunnel axis $\alpha > 30°$, with dip angle of $\beta > 75°$	Other combination
K_2	0.4-0.6	0-0.2	0.2-0.4

The revision coefficient influenced by the natural stress (K_3)   Table 4-19

Natural stress \ BQ	<250	250-350	350-450	450-550	>550
Extremely high stress area	1.0	1.0-1.5	1.0-1.5	1.0	1.0
High stress area	0.5-1.0	0.5-1.0	0.5	0.5	0.5

Notes: The extremely high stress area refers to $\sigma_{cw}/\sigma_{max} < 4$, high stress area refers to $\sigma_{cw}/\sigma_{max} = 4\text{-}7$, and σ_{max} is the maximum natural stress in the plane perpendicular to cavern axis.

The groundwater influencing factor K_1 and the main structural plane revision parameter K_2 can be obtained through the statistics of joints and fissures. Through the measurement results of in-situ stress, the revision coefficient influenced by the natural in-situ stress K_3 can be obtained. Combining with the above equation, the obtained [BQ] value is shown in Table 4-20.

The parameters of each tunnel and [BQ] value results   Table 4-20

Position	Basic quality index of rock mass (BQ)	Revision [BQ]	Qualitative level
−780 m horizontal measuring point 1	475.25	295.25	IV
−780 m horizontal measuring point 2	540.00	360	III
−795 m horizontal measuring point 1	498.75	318.75	IV
−795 m horizontal measuring point 2	503.75	323.75	IV

continued

Position	Basic quality index of rock mass(BQ)	Revision [BQ]	Qualitative level
−825 m horizontal measuring point 1	545.25	365.25	Ⅲ
−825 m horizontal measuring point 2	485.50	305.5	Ⅳ
−825 m horizontal measuring point 3	538.25	358.25	Ⅲ
−825 m horizontal measuring point 4	504.75	324.75	Ⅳ
−870 m horizontal	533.25	353.25	Ⅲ
−885 m horizontal	460.25	300.25	Ⅳ
−900 m horizontal	463.00	303.00	Ⅳ
−915 m horizontal measuring point 1	477.35	317.35	Ⅳ
−915 m horizontal measuring point 2	484.35	324.35	Ⅳ

4.4.2 Rock mass rating classification

The rock mass rating classification of rock mass uses the total score *RMR* of rock compressive strength, *RQD*, joint interval, weathered and metamorphic features, joint continuity and backfilling, and strike and tilt of groundwater joint indexes to determine the rock mass grade. Select *RMR* method to implement quality evaluation of the deep rock mass in the Sanshandao Gold Mine. Conduct the in-situ stress revision by *RMR* classification method on this basis, considering the in-situ stress impact on the stability of rock mass, and establish the *IRMR* classification index. Use *RMR* and evaluation system of rock mass *IRMR* to evaluate the quality and stability of rock mass at each deep stope in the Sanshandao Gold Mine, in order to provide fundamental basis to optimize the deep mining method and the structural parameters of stope.

The *RMR* evaluation system includes six evaluation indexes: R_1 is rock compressive strength, R_2 is rock quality index *RQD*, R_3 is joint interval, R_4 is joint state, R_5 is groundwater state, R_6 is the revision parameter of joint direction influencing engineering. The *IRMR* also increases R_7 for revision parameter of in-situ stress. Sum up the above indexes scores of each parameter to get the values of the rock mass are as shown in equations (4-5) and (4-6):

$$RMR = R_1 + R_2 + R_3 + R_4 + R_5 + R_6 \qquad (4-5)$$
$$IRMR = R_1 + R_2 + R_3 + R_4 + R_5 + R_6 + R_7 \qquad (4-6)$$

According to the geological investigation and rock mechanics research data of

4 Rock Mass Investigation and Rock Quality Evaluation

Sanshandao, the scores of rock mass quality indexes of each middle section of the deep and main stope based on RMR evaluation system are calculated as follows:

(1) The rock compressive strength

the sampling rock core and stope rock sample through in-situ stress measurement, conduct mechanical properties tests on the -780 m middle section, -795 m segment, -825 m middle section, -870 m middle section, -885 m segment, -900 m segment, and -915 m middle section. Refer to previous research experiences and data on mines, determine the tensile strength of rock at each middle section of the deep as shown in Table 4-21.

The tensile strength of rock at each middle section   Table 4-21

Depth	Compressive strength(MPa)
-780 m middle section	80.23
-795 m segment	79.38
-825 m middle section	82.74
-870 m middle section	76.04
-885 m segment	88.90
-900 m segment	87.30
-915 m middle section	75.76

According to the data in the above table, the score of R_1 (Table 4-22) can be calculated with equation (4-7):

$$R_1 = -0.0003\sigma_{ucs}^2 + 0.135\sigma_{ucs} + 0.9023 \tag{4-7}$$

The score of rock uniaxial compressive strength R_1 Table 4-22

Engineering position	-780 m middle section	-795 m segment	-825 m middle section	-870 m middle section	-885 m segment	-900 m segment	-915 m middle section
R_1 score	9.8	9.7	10.0	9.4	10.1	10.4	9.3

(2) RQD value

The RQD value is determined by selecting the rock core of the drilling corresponding to the investigation site. Through statistical analysis of drilling data related to each middle section prospecting line, the RQD value of each engineering position is obtained. The RQD score is obtained according to equation (4-8), as shown in Table 4-23.

$$R_2 = 0.0012I_{RQD}^2 + 0.0692I_{RQD} + 1.23 \tag{4-8}$$

RQD value score   Table 4-23

Engineering position	−780 m middle section	−795 m segment	−825 m middle section	−870 m middle section	−885 m segment	−900 m segment	−915 m middle section
RQD(%)	87.4	91.7	75.9	22.5	82.1	82.6	64.6
score	16.4	17.6	13.4	3.4	15.0	15.1	10.7

(3) The joint interval

According to the statistical analysis results of deep joint investigation, the average joint interval is used for evaluation. The average joint interval is substituted into equation (4-9) to calculate the score of joint interval R_3, as shown in Table 4-24.

$$R_3 = 3.5411 \ln J_v + 16.6617 \qquad (4-9)$$

Joint interval score   Table 4-24

Engineering position	Joint interval(cm)	Score
−780 m middle section measuring point 1	26.4	11.9
−780 m middle section measuring point 2	27.1	12.0
−795 m segment measuring point 1	28.1	12.2
−795 m segment measuring point 2	28.3	12.2
−825 m middle section measuring point 1	33.6	12.8
−825 m middle section measuring point 2	30.3	12.4
−825 m middle section measuring point 3	32.1	12.8
−825 m middle section measuring point 4	30.5	12.4
−870 m middle section	28.2	12.2
−885 m segment	30.8	12.5
−900 m segment	32.9	12.8
−915 m middle section measuring point 1	23.3	11.5
−915 m middle section measuring point 2	23.4	11.5

(4) The structural plane condition

According to the site observation, combining with the actual geological engineering conditions of deep joints and fissures in Sanshandao, and with the reference to scoring standard of structural plane conditions integrally, the RMR scores of structural plane conditions are obtained, as shown in Table 4-25.

Table of joint surface condition score   Table 4-25

Position of measuring point	Score
−780 m middle section measuring point 1	17
−780 m middle section measuring point 2	17
−795 m segment measuring point 1	20
−795 m segment measuring point 2	19
−825 m middle section measuring point 1	22
−825 m middle section measuring point 2	21
−825 m middle section measuring point 3	20
−825 m middle section measuring point 4	20
−870 m middle section middle section	18
−885 m segment	20
−900 m segment	19
−915 m middle section measuring point 1	17
−915 m middle section measuring point 2	17

(5) Groundwater

Through the investigation of groundwater in the Sanshandao mining area, the groundwater condition of the deep transportation tunnels at each middle section and the main stope are counted, and the scores of groundwater are obtained as shown in Table 4-26.

Table of groundwater condition score   Table 4-26

Engineering position	−780 m middle section	−795 m segment	−825 m middle section	−870 m middle section	−885 m segment	−900 m segment	−915 m middle section
Groundwater condition	Severe seepage	Humid seepage	Humid seepage	Relatively humid	Seepage and dripping	Humid seepage	Severe seepage
Score	0	7	7	12	4	7	0

(6) The revision of joints and fissures strike

The ore body of Sanshandao Gold Mine has a general strike of 40°, a local strike of 70°-80°, SE trend, with the dip angle of 45°-75°. The long axis of the stope is laid out along the strike or perpendicular to the strike, and the excavation direction of the cross gateway is 124°. The revision score of R6 can be obtained according to the strike and dip angle of the hanging wall and footwall of rock mass, ore body and the main joints and fissures at each middle section, as shown in Table 4-27.

The revision score table of joints and fissures strike　　Table 4-27

Engineering position	−780 m middle section	−795 m segment	−825 m middle section	−870 m middle section	−885 m segment	−900 m segment	−915 m middle section
The advantage of joint strike to the development and excavation	Ordinary	Ordinary	Ordinary	Ordinary	Favorable	Adverse	Ordinary
Score	−5	−5	−5	−5	0	−10	−5

(7) The revision of in-situ stress

According to the results of deep in-situ stress measurement, using the regression equation of maximum horizontal principal stress, minimum horizontal principal stress and vertical principal stress in the deep in-situ stress field of Sanshandao, the in-situ stress values of each middle section of the deep and stope can be calculated as shown in Table 4-28 below. The revision of in-situ stress can refer to the revision of in-situ stress in the *Standard for Classification of Engineering Rock Mass*, which has been proposed in China in 2015 for the classification of underground cavern rock mass, as shown in Table 4-29.

Table of in-stress value at each middle section　　Table 4-28

Engineering position	Maximum horizontal principal stress σ_{hmax} (MPa)	Vertical principal stress σ_z (MPa)	Minimum horizontal principal stress σ_{hmix} (MPa)
−780 m middle section	34.714	21.104	20.749
−795 m segment	35.335	21.494	21.139
−825 m middle section	36.595	22.274	21.919
−870 m middle section	38.485	23.444	23.089
−885 m segment	39.115	23.834	23.479
−900 m segment	39.745	24.224	23.869
−915 m middle section	40.375	24.614	24.259

The revision coefficient influencing in-situ stress　　Table 4-29

R_7	State of in-situ stress	Extremely high in-situ stress	High in-situ stress	Low in-situ stress
	Score	−15	−10	0

Notes: The extremely high stress refers to $R_c/\sigma_{max} < 4$, high stress refers to $R_c/\sigma_{max} < 4-7$, and low stress refers to $R_c/\sigma_{max} > 7$, σ_{max} is the maximum natural stress in the plane perpendicular to cavern axis.

■ 4 Rock Mass Investigation and Rock Quality Evaluation ■

The ratio of rock compressive strength to the maximum horizontal principal stress is defined as the rock mass damage coefficient Z. In view of the "leaping" standard for the score of the in-situ stress revision, the in-situ stress revision in Table 4-30 is revised by continuous scoring method in connection with the rock mass damage coefficient Z and the score, as shown in Table 4-31.

The revision of rock mass quality score considering the influence of in-situ stress   Table 4-30

R_7	Z	1-2	2-3	3-4	4-5	5-6	6-7	>7
	Score	−15	−12	−9	−7	−5	−3	0

The regression fitting is conducted between Z value and score in Table 4-30 to obtain a continuous scoring equation between R_7 and its score, whose function relationship is shown in equation (4-10). Also, the revised score of in-situ stress revision coefficient R_7 on $IRMR$ is shown in Table 4-31.

$$R_7 = 3Z - 18 \qquad (4\text{-}10)$$

The rock mass damage coefficient Z value   Table 4-31

Engineering position	Middle section of −780 m	Segment of −795 m	Middle section of −825 m	Middle section of −870 m	Segment of −885 m	Segment of −900 m	Middle section of −915 m
Damage coefficient	2.850	4.552	4.552	2.850	2.850	2.919	2.919
Score	−9.450	−4.345	−4.345	−9.450	−9.450	−9.243	−9.243

According to the RMR quality evaluation indexes of rock mass shown in Table 4-32, the features of RMR and $IRMR$ quality evaluation indexes are obtained through site investigations and laboratory tests, and the quality evaluation results of Sanshandao Gold Mine rock mass are obtained through calculation, as shown in Table 4-33.

The quality evaluation indexes of Sanshandao rock mass   Table 4-32

Position of measuring point	Evaluation index						
	R_1	R_2	R_3	R_4	R_5	R_6	R_7
−780 m middle section measuring point 1	9.8	16.4	11.9	17	0	−5	−4.3
−780 m middle section measuring point 2	9.8	16.4	12.0	17	0	−5	−4.3

continued

Position of measuring point	Evaluation index						
	R_1	R_2	R_3	R_4	R_5	R_6	R_7
−795 m segment measuring point 1	9.7	17.6	12.2	20	7	−5	−4.3
−795 m segment measuring point 2	9.7	17.6	12.2	19	7	−5	−4.2
−825 m middle section measuring point 1	10.0	13.4	12.8	22	7	−5	−9.5
−825 m middle section measuring point 2	10.0	13.4	12.4	21	7	−5	−9.5
−825 m middle section measuring point 3	10.0	13.4	12.8	20	7	−5	−8.2
−825 m middle section measuring point 4	10.0	13.4	12.4	20	7	−5	−9.0
−870 m middle section	9.4	3.4	12.2	18	12	−5	−9.5
−885 m segment	10.1	15.0	12.5	20	4	0	−9.5
−900 m segment	10.4	15.1	12.7	19	7	−10	−9.2
−915 m middle section measuring point 1	9.3	10.7	11.5	16	0	−5	−9.2
−915 m middle section measuring point 2	9.3	10.7	11.5	17	1	−6	−9.2

The grade and quality evaluation of rock mass determined by total score of rock mass rating classification   Table 4-33

Engineering position	RMR total score	IRMR total score	Grade of rock mass	Description of quality
−780 m middle section measuring point 1	50.1	45.8	Ⅲ	Ordinary
−780 m middle section measuring point 2	50.2	45.9	Ⅲ	Ordinary
−795 m segment measuring point 1	61.5	57.2	Ⅲ	Ordinary
−795 m segment measuring point 2	60.5	56.3	Ⅲ	Ordinary
−825 m middle section measuring point 1	60.2	50.7	Ⅲ	Ordinary

Engineering position	RMR total score	IRMR total score	Grade of rock mass	Description of quality
−825 m middle section measuring point 2	58.8	49.3	III	Ordinary
−825 m middle section measuring point 3	58.2	50	III	Ordinary
−825 m middle section measuring point 4	57.8	48.8	III	Ordinary
−870 m middle section	50	40.5	IV	Poor
−885 m segment	61.6	52.1	III	Ordinary
−900 m segment	54.2	45	III	Ordinary
−915 m middle section measuring point 1	42.5	33.3	IV	Poor
−915 m middle section measuring point 2	43.5	34.3	IV	Poor

4.4.3 Q-system classification

Barton Quality (Q) Classification of Rock Mass, also known as the Q System, is the most widely used rock mass quality classification method. In 1974, Nick Barton et al in Norway have established this classification method on the basis of 212 tunnel engineering examples. It mainly considers the influence of rock mass intactness, joint characteristics, groundwater and in-situ stress, and determines the Q value of rock mass quality index reflecting the stability of tunnel surrounding rock with six parameters:

$$Q = \frac{RQD}{J_n} \cdot \frac{J_r}{J_a} \cdot \frac{J_w}{SRF} \tag{4-11}$$

where　RQD—rock quality index;
　　　　J_n—Qty. of joint set;
　　　　J_r—joint roughness coefficient;
　　　　J_a—joint alteration coefficient;
　　　　J_w—joint-water reduction coefficient;
　　　　SRF—stress reduction coefficient.

The combination of the six parameters in equation (4-11) reflects the three aspects of rock mass quality, namely RQD/J_n is the intactness of rock mass; J_r/J_a is the morphology of structural plane, backfilling features and its secondary changes; J_w/SRF represents the influence of water and the existed other stresses on rock mass quality.

During classification, according to the measured data of the six parameters, check the table to determine their respective values, which are substituted equation (4-11) to obtain the Q value of rock mass. Q ranges from 0.001 to 1000, representing the quality of surrounding rock from extremely poor extrusive rock to the excellent hard intact rock, which is divided into 9 quality grades, as shown in Table 4-34.

Table of rock mass classification grade of Q system　　　　Table 4-34

Q value	0.001-0.01	0.01-0.1	0.1-1	1-4	4-10	10-40	40-100	100-400	400-1000
Grade	9	8	7	6	5	4	3	2	1
Description	Worst	Extremely poor	Very bad	Poor	Ordinary	Good	Very good	Extremely good	Excellent

In the Q system, the parameters of J_n, J_r, J_a, J_w and SRF are selected according to Table 4-35 to Table 4-39.

The J_r value of joint roughness coefficient　　　　Table 4-35

Joint roughness coefficient	J_r value
(A) Direct contact with joint wall (without mineral backfilling, or only with thin layer of mineral backfilling)	
(B) Direct contact with 10 cm dislocation of front joint wall (with thin layer of mineral backfilling)	
a. Discontinuous joint	4
b. Rough or irregular, undulate	3
c. Smooth, undulate	2
d. Smooth, undulate	1.5
e. Rough or irregular, straight	1.5
f. Smooth, straight	1.0
g. Glossy, straight	0.5
(C) Indirect contact with joint wall at dislocation (with thick layer of mineral backfilling)	
h. With the clay zone having the thickness to prevent the contact with joint wall	1.0

4 Rock Mass Investigation and Rock Quality Evaluation

continued

Joint roughness coefficient	J_r value
i. With the sand, gravel, and shattered zone having the thickness to prevent the contact with joint wall	1.0

The J_n value of quantity of joint set   Table 4-36

Quantity of joint set	J_n value
a. Block, no or few joint	0.5-1
b. One set of joint	2
c. One set of joint with random joint	3
d. Two sets of joints	4
e. Two sets of joints with random joint	6
f. Three sets of joints	9
g. Three sets of joints with random joint	12
h. Joints over four sets, with severe joints, rock fragmentation	15
i. Shattered rock, similar to soil	20

The J_w value of joint-water reduction coefficient   Table 4-37

	Hydraulic pressure (MPa)	J_w value
a. Dry at excavation, or with small flow locally (<5 L/min)	<0.1	1.0
b. Medium flow or with medium pressure, occasionally outrushing with backfilling	0.1-0.25	0.66
c. Large flow or high pressure in the hard rock containing un-backfilled joint	0.25-1	0.5
d. Large flow or high hydraulic pressure, attenuation with time	0.25-1	0.33
e. Extra large flow or high hydraulic pressure, attenuation with time	>1	0.2-0.1
f. Extra large flow or high hydraulic pressure, not attenuation with time	>1	0.1-0.05

The J_a value of joint weathered alteration coefficient   Table 4-38

	Residual angle of friction(°)	J_a value
(A) Direct contact with joint wall (without mineral backfilling, or only with thin film coverage)		
a. Tightly closed, hard, not softening, water-proof backfilling, such as quartz, epidote	—	0.75

continued

	Residual angle of friction(°)	J_a value
b. Unmetamorphosed joint wall, only with stain on the surface	25-35	1.0
c. Slightly metamorphic joint wall, without softening mineral coverage, sand grain, loose clay, and other backfilling	25-30	2.0
d. With silty or sand thin film coverage, with few clay component (without softening)	20-25	3.0
e. Softening or low friction clay mineral coverage(such as kaolinite, mica, nitrite, steatite, plaster, graphite, few expansive clay, etc.)	8-16	4.0
(B) Direct contact with 10 cm dislocation of front joint wall (with thin layer of mineral backfilling)		
f. Fissures containing sand grain, loose clay, etc.	25-30	4.0
g. Intensively extra consolidated, not softening clay mineral backfilling(continuous, but the thickness less than 5 mm)	16-24	6.0
h. Medium or slightly extra consolidated, clay backfilling consisted of softening mineral(with the thickness less than 5 mm)	8-12	8
i. Expansive clay backfilling (continuous, with the thickness less than 5 mm) such as montmorillonite, kaolinite, etc., J_a depends on the content of expansive clay particle and inlet of water, etc.	6-12	8-12
(C) Indirect contact with joint wall at dislocation (with thick layer of mineral backfilling)		
KLM incomplete or broken rock and clay strip area (for clay condition, refer to G. H. J)	6-24	6,8 or 8-12
N silty or sand clay strip area, containing few clay component (not softening)	—	5.0
OPR thickness continuous area or clay strip (for clay condition, refer to G. H. J)	6-24	10,13 or 13-20

The *SRF* value of stress reduction coefficient   Table 4-39

	SRF value
(A) The weak zone cross with the excavation direction, it will lead to rock mass relaxing at excavation	
a. Weak zone of incomplete rock containing clay or chemical weathered occurs repeatedly and the surrounding rock is very loose (at any depth)	10.0
b. Single weak zone of incomplete rock containing clay or chemical weathered occurs repeatedly (at the excavation depth ≤50 m)	5.0

continued

			SRF value
c. Single weak zone of incomplete rock containing clay or chemical weathered occurs repeatedly(at the excavation depth >50 m)			2.5
d. Multiple shear zones in the hard rock(without clay), the surrounding rock is loose(at any depth)			7.5
e. Single shear zone in the hard rock(without clay), the surrounding rock is loose(at the excavation depth ≤50 m)			5.0
f. Single shear zone in the hard rock(without clay), the surrounding rock is loose(at the excavation depth >50 m)			2.5
g. Loose and opening joints, severe joints or in small blocks, etc. (at any depth)			5.0
(B) Hard rock, issue of rock stress	σ_c/σ_1	σ_θ/σ_c	SRF value
h. Low stress, adjacent to surface, opening joint	>200	<0.01	2.5
i. Medium stress, most favorable stress condition	200-10	0.01-0.3	1.0
j. High stress, extremely tight structure, ordinary beneficial to stability, also might not suitable for the stability of tunnel side wall	10-5	0.3-0.4	0.5-2.0
k. Medium plate crack occurs after 1 hour in block rock mass	5-3	0.5-0.65	5-50
l. Plate crack and rockburst occur within several minutes in block rock mass	3-2	0.65-1	50-200
m. Severe rockburst occurs in block rock mass(the burst in of strain and direct dynamic deformation)	<2	>1	200-400
(C) Extruding rock: plastic flow of soft rock under the influence of high stress		σ_θ/σ_c	SRF value
n. Slightly extruding rock stress		1-5	5-10
o. Severely extruding rock stress		>5	10-20
(D) Expansive rock, rock will occur chemical expansion due to the existence of water			SRF value
p. Slightly expansive rock stress			5-10
q. Severely expansive rock stress			10-15

According to the classification parameters of quality evaluation as shown in Table 4-35 to Table 4-39, the Q system quality evaluation index is obtained, and the Q system evaluation results of rock mass quality in the Sanshandao Gold Mine are obtained through calculation, as shown in Table 4-40.

The Q system parameter values and evaluation results   Table 4-40

Evaluation index	−780 m middle section	−795 m segment	−825 m middle section	−870 m middle section	−885 m segment	−900 m segment	−915 m middle section
RQD(%)	87.4	91.7	75.9	22.5	82.1	82.6	64.6
J_n	7	7	9	7	8	10	9
J_r	2	1.5	2	1.5	1	1.5	1.5
J_a	4	4	4	4	4	4	4
J_w	0.65	0.6	0.6	0.65	0.58	0.55	0.58
SRF	1	1.5	2	2	2	2	2
Q value	2.86	28.21	5.55	0.41	7.94	10	0.02
Classification of rock mass	6	4	5	7	4	4	7
Description of quality	Poor	Good	Ordinary	Very bad	Good	Good	Extremely poor

4.4.4 The comprehensive evaluation of deep rock mass quality

The Engineering Rock Mass Classification of the National Standard (BQ) is a method combining qualitative and quantitative analyses, experience judgment and test calculation. The qualitative classification only requires site investigation, and quantitative classification only requires tests of uniaxial compressive strength, rock mass and elastic wave velocity of rock block.

Three basic parameters of rock mass rock mass rating classification (RMR) are quantitative and the other three basic parameters are qualitative. It is a semi-quantitative and semi-qualitative method, which requires the participation of experienced geologists.

The Barton Quality Classification (Q system) except the rock quality index RQD, the other 5 indexes are obtained based on the description table of the site investigation. It is basically a qualitative classification method, which requires the participation of experienced geologists, with larger subjective arbitrariness.

There is a linear relationship between rock mass rating classification (RMR) and Quality Index of Engineering Rock Mass (BQ) value, and their relationship is: [BQ] =0.089RMR+21.378.

The Grade I of [BQ] is equivalent to above Grade II to Grade I of RMR, Grade V of [BQ] is equivalent to RMR Ⅳ, for all the rest grades, they are differed about half a grade, for more details, please refer to Table 4-41.

The relationship table of BQ classification and RMR classification   Table 4-41

[BQ] grade	[BQ] value	RMR value	RMR grade
I	>550	>70	II_{above}-I
II	450-550	60-70	II_{below}
III	350-450	50-60	III_{above}
IV	250-350	40-50	III_{below}
V	<250	<40	IV-V

The relationship between Barton Quality Classification (Q system) and Quality Index of Engineering Rock Mass [BQ] value is shown in Table 4-42. From the table, it can be seen that the corresponding relationship between the national standard classification and Q system classification is good. Grade I of the national standard is equivalent to above upper Grade 2 of Q system, which is the extremely good and excellent rock mass. Grade V of the national standard is equivalent to Grade 6 and below Grade 6 of Q system, which is the bad to the worst rock mass. For all the rest grades, they are differed about half a grade.

The relationship table of BQ classification and Q system   Table 4-42

[BQ] grade	[BQ] value	Q value	Grade of Q system
I	>550	>290	2_{above}-1
II	450-550	70-290	3_{above}-2_{below}
III	350-450	16-70	4_{above}-3_{below}
IV	250-350	4-16	5-4_{below}
V	<250	<4	6-9

The above three methods are used to carry out quality classification evaluation of deep rock mass in the Sanshandao Gold Mine, and the results obtained are shown in Table 4-43. According to the results, BQ method classification result for -780 m to -825 m horizontal rock mass is Grade III on average, for -825 m and -915 m horizontal are Grade IV, with poor stability; RMR method result shows the majority of deep rock mass quality in the Sanshandao Gold Mine is Grade III and Grade IV, and rock mass quality of -915 m horizontal tunnel is Grade IV. The Q system classification result shows the relatively poor quality of deep rock mass in the gold mine, which belongs to relatively poor stability of rock mass. Thus it requires special support reinforcement, and conducts real-time monitor on the activities of confining pressure in the process of mining and production.

The list of classification of deep rock mass quality　　　　Table 4-43

Evaluation index	−780 m middle section	−795 m middle segment	−825 m middle section	−870 m middle section	−885 m middle segment	−900 m middle segment	−915 m middle section
[BQ] classification	IV	III	IV	IV	III	III	IV
RMR classification	III	III	III	IV	III	III	IV
Description of quality	Ordinary	Ordinary	Ordinary	Poor	Ordinary	Ordinary	Poor
Q classification	6	4	5	7	4	4	7
Description of quality	Poor	Good	Ordinary	Extremely poor	Good	Good	Extremely poor

The results of the three methods of BQ classification, RMR classification and Q system are slightly different, and the reasons for the differences are mainly as follows:

(1) The three classification methods have different participating factors and emphases in classification;

(2) There is an error between the parameters calculated by classification and the reality;

(3) The selection of non-quantitative parameters is affected by human subjective factors.

Although the applications and emphases of the main rock mass quality classification methods are different, their purposes are to reflect the complexity of engineering rock mass structure, and provide references for the evaluation of engineering rock mass stability and comprehensive utilization of rock mass. The above three types of rock mass quality evaluation methods have summarized the factors that control the development features of rock mass quality as the structural intactness of rock mass, rock mass strength and other geological and engineering factors, including geological factors as the leading factor, which provides the basis for conversion and comparison among various evaluation classification methods.

According to the conversion and comparison principle of the above three classification methods (Table 4-41, Table 4-42), the classification results of rock mass are compared and converted, and the classification results of BQ classification, RMR classification and Q system are approximately consistent. This indicates that the classification results of deep rock mass quality in the Sanshandao Gold Mine are accurate and reliable.

5

Rock Chemical Damage Induced by Deep Hydrochemical Environment

5.1 Analysis of deep hydrochemical environment

The mining area of Sanshandao Gold Mine is situated in the sea coast of Laizhou Bay where the geological structure is more complex, and the brine of bedrock is the ancient seawater. Its supplied water sources are classified into the HCO_3-Ca and Cl-Na types according to the hydrological characteristics. With the increase of mining depth, the seepage and water gushing phenomena of surrounding rocks caused by the shafting and drifting engineering development and mining operation are gradually appeared. As shown in Figure 5-1, the water gushing and water inrush phenomena of tunnel roof and two sides' surrounding rocks in some deep sections of mining area is highly significant, thereby result in a large area of water accumulation on the floor, and cause a negative influence on the normal safety production activities of mine.

In order to ensure the safety and high efficiency of deep mining activities, it is necessary to consider the influence of groundwater, monitor the supplied water inflow of mine area, and reveal the dynamic rule of water inflow in the mine pit. The main supplied water body of Sanshandao Gold Mine is the ancient seawater. The groundwater is composed of Quaternary pore water and bedrock fissure water, while the water-quality constituents are more complex. Therefore, it has a great significance to analyze the water chemical composition of groundwater in the mining area for the prevention and control of water inrush and water gushing disasters and the discrimination of seawater gushing channel.

5.1.1 Dynamic monitoring and characteristic analysis of water gushing

In order to master the spatial and temporal distribution rule of water gushing volume in the mining area, and accumulate the dynamic multi-period observation data of groundwater, the 12 monitoring points with the representative character-

■ 5 Rock Chemical Damage Induced by Deep Hydrochemical Environment ■

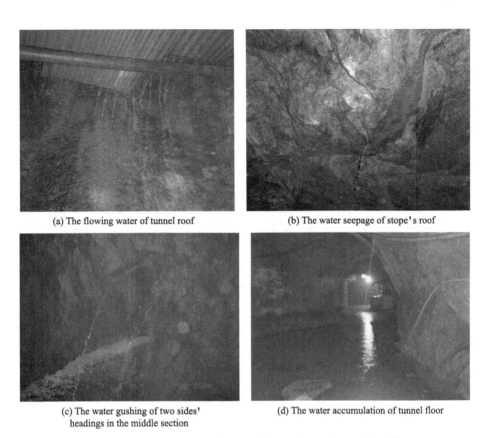

(a) The flowing water of tunnel roof

(b) The water seepage of stope's roof

(c) The water gushing of two sides' headings in the middle section

(d) The water accumulation of tunnel floor

Figure 5-1　The water gushing problem of groundwater faced by the deep mining in the mining area under Sanshandao Gold Mine

istics of water gushing in the mining area are established to observe for the tunnel systems of 3 middle sections in the Sanshandao mining area on the basis of hydrogeological survey of mining area and according to the survey situation of water point in each deep middle section. It can be seen as Table 5-1 for the position of each monitoring point of water gushing volume and the statistics of water gushing characteristics. Considering that the water gushing of deep mine pit under the mining area is mainly composed of cluster seepage, and the seepage point with the largest water drop volume and relatively stable flow is selected as the monitoring point for the observation of its water flow, one measuring cup of 1000 mL is used for the observation to coordinate with a stopwatch, and then the number of seepage points in the seepage area is counted, so that the water gushing volume of each seepage area can be worked out. Through the development of multi-period monitoring works, a large number of observation data are obtained, and provide the foundation and basis for mastering the hydrogeological condition of each middle section in the mining area and revealing the dynamic rule of water gushing volume in the mine pit.

The monitoring point position of water gushing in the mining area under Sanshandao Gold Mine and the characteristics of water gushing　　Table 5-1

Middle Section	Monitoring Point No.	Position	Characteristic of Seepage
Middle Section in −510 m Level	510-1	around the Line 1480 of main tunnel in the middle section	Seepage of roof fracture
	510-2	around the Line 1600 of main tunnel in the middle section	Effluent of drilling hole in the right side of roof
	510-3	located in the F3 fault of main tunnel in the middle section	Flowing water from the roof and both sides of right zone
Middle Section in −555 m Level	555-1	located in the accurate mining tunnel of Line 1560 in the north tunnel	Seepage of fracture in the left zone
	555-2	around the return air shaft of Line 1720 in the north tunnel	Flowing water from the roof fracture
	555-3	located in the F3 fault of main tunnel in the middle section	Water gushing from the roof and right zone
	555-4	located in the mining combination of NO. 10 in S12186 stope	Flowing water from the fracture of right zone
Middle Section in −600 m Level	600-1	located in the entry of water pump house in the north tunnel	Flowing water from the fracture of right side in the roof
	600-2	located in S13165 stope	Seepage of roof fracture
	600-3	around the 1680 line of north tunnel	Water gushing from the drilling hole of right zone
	600-4	located in the F3 fault of main tunnel in the middle section	Water gushing from the roof and both zones
	600-5	located in the entry of No. 1 mining combination in S13186 stope	Flowing water from the fractures of roof and left zone

Analyze the data of water gushing volume obtained from the field observation in the monitoring point of each mid-section. By taking the monitoring time as the horizontal axis and the water flow as the vertical axis, draw the curve of water flow with the change of time in each deep monitoring point of Sanshandao

5 Rock Chemical Damage Induced by Deep Hydrochemical Environment

Gold Mine, which is shown as Figure 5-2. Analyze the changing situation and characteristics of water flow in the monitoring point of each middle section.

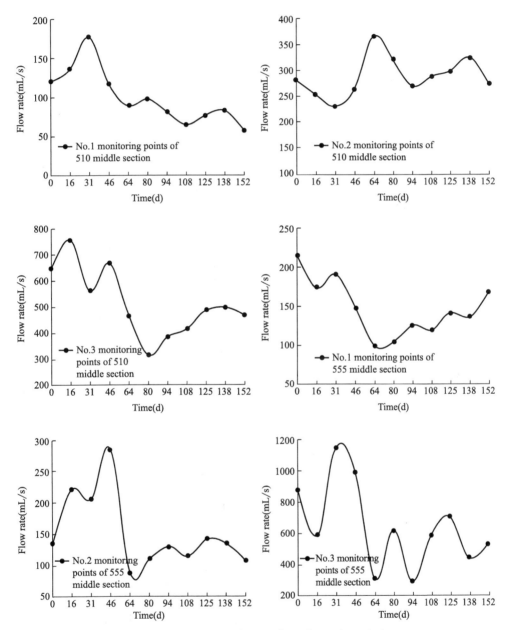

Figure 5-2 The curve of water flow change in each deep monitoring point of water gushing in the mining area (one)

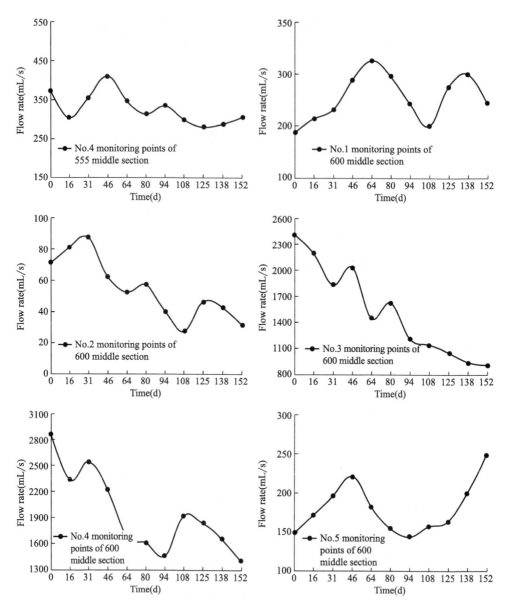

Figure 5-2　The curve of water flow change in each deep monitoring point of water gushing in the mining area (two)

The deep mining area of Sanshandao Gold Mine is under the high-temperature, high-salinity and acidic conditions, and the erosion phenomenon of surrounding rock is more serious, which are shown as Figure 5-3. As shown in the pictures, the main components of white crystalline material do not only contain the sodium chloride, but also contain a small amount of kaolin mineral composition precipitated out after the dissolution of feldspar mineral through the diffraction analysis of XRD. The red crystalline mineral is the ferric hydroxide generated from the oxidation of pyrite under the condition of water.

 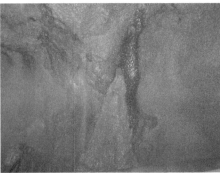

Figure 5-3 The erosion phenomenon of deep surrounding rock in Sanshandao Gold Mine

3 monitoring points are totally established in the middle section of -510 m Level, which are respectively located in Line 1480 and Line 1600 as well as the F3 fault of main tunnel in the middle section. The characteristic of water gushing in No. 1 monitoring point presents as the seepage of fracture. The water flow is smaller and more stable, while the fluctuation is not large. The water gushing volume of No. 2 monitoring point appears a certain range of fluctuation in the middle of monitoring period, and tends to be stable in the later period. It is 289 mL/s for the average flow rate during the whole monitoring period. Due to locating in the water diversion channel of F3 fault, the water gushing volume of No. 3 monitoring point during the monitoring period is over 300 mL/s. The water gushing volume of No. 3 monitoring point is larger. It can be seen from the monitoring data of this observation point that the water flow has a tendency of decrease. The reason of appeared phenomenon may be that the shotcreting support operation nearby blocks a part of channel of groundwater migration, also may be related to that the production activity of middle section above the level of this monitoring point is reduced and the production water consumptions of filling etc is lowered down.

4 monitoring points are totally established in the middle section of -510 m level. The monitoring points of No. 1 and No. 2 mainly are the partial seepage and water flow. Although the water gushing volume of No. 2 monitoring point appears a large fluctuation during the monitoring period, the water gushing volumes of both monitoring points are not large. The monitoring point of No. 3 is close neighbor to the F3 fault, and the water gushing volume is bigger. Its high-

est water gushing volume during the monitoring period is 1140 mL/s. The water gushing volume of this monitoring point tends to be stable at the later period of monitoring. However, its overall flow is still at a high level. The monitoring point of No. 4 is located in the north region of F3 fault. Since the fracture of rock mass is caused by the influence of development engineering, and the shotcreting construction is not carried out in time, an effluent point is formed. The average water gushing volume of this monitoring point is 329 mL/s, and relatively stable in the whole. Since there is not any significant production activity around this measuring point during the monitoring period, the water gushing volume of this monitoring point will be likely increased by the filling water of later mining operation. At that time, it will be necessary to make the further observation for this monitoring point. On the whole, the water gushing volume of middle section in -555 m level is obviously increased when comparing with that middle section in -510 m level.

Since there are more positions of seepage and water gushing in the middle section of -600 m level, 5 monitoring points are totally set. The monitoring points of No. 1 and No. 2 are located in the south region of north tunnel, while the monitoring flows is not large and appears a decreased tendency. Due to the exploring construction, the channel of groundwater migration is broken through, and a mass flow of spray water gushing is formed in the surrounding rocks of right zone in the tunnel of No. 3 monitoring point. The maximum volume of water gushing reaches to 2405 mL/s. Since the producing activities in the middle section of -600 m level during the monitoring period are mainly the accurate mining and exploration, and the mining work is not yet carried out on a large scale, the enough attention is not yet attracted to the water gushing situation of this water gushing point, and any effective plugging measure is not taken. Only a simple treatment of water plugging has been made in the later period, so that the water gushing volume is declined to some extent. The monitoring point of No. 4 is located in F3 fault, while the hydrogeological condition is bad, and the water gushing volume is very large. This causes a large area of water accumulation in the floor of transportation heading. Based on this situation, the water sump and water pump house are established in the middle section of -600 m level in the mining area to dredge the large area of water gushing. The monitoring point of No. 5 is located in the north of F3 fault region. Influenced by the disturbance of surrounding development projects, the observed flow appears an increased tendency. Viewing from the observed data, the average flow rate of 5 monitoring points in the middle section of -600 m level reaches to 794. 6 mL/s. It is observably bigger than that of middle sections in -510 m level and -555 m level.

The monitoring result of water gushing volume shows that: The water volume of monitoring points in the northern region of main tunnel in each middle section is generally greater than that in the southern region. In the north of F3 fault and the west of F1 fault, the broken alteration zones are widely distributed, and intensify the occurrence of water gushing and inrush disasters in the mining area to a certain extent. With the progress of mining work, the underground goafs are increased, and result in the continuous destruction of surrounding rock mass. Since the destruction of rock mass maybe results in the connection of some fractures, and maybe make the original connection of fractures blocked and closed too, the flow situation in the same seepage position will be led to continuously change with time. The mining area under Sanshandao Gold Mine is close to the coast, and belongs to the mining of coast. In addition, there exists the joint development in the north of mining area and the water diversion of F3 fault zone, so that the potential disasters of water gushing and inrush make a great threat to the underground engineering. In order to effectively prevent the flood disaster, it shall be continuously strengthened for the observation on the point location where there is the serious water seepage and the flow appears an increasing tendency. For the key area where there is a large volume of water gushing, and the safety production may be affected, the plugging measures of shotcreting construction etc can be taken if necessary, so as to eliminate any hidden danger.

5.1.2 Hydrochemical composition and content analysis of underground water

Unique hydrochemical characteristics take form in natural water during its long-term interaction with the surrounding environment, and to a certain extent it takes records of such water cycle information as conditions for existence of water body, seepage path and source of replenishment. Using content and ratio of chemical components of the water body helps find out environment and hydrodynamic conditions in which it is located so as to provide basis for probing into origin of causes for underground water, source of replenishment and for judging its hydraulic relationship with other water bodies.

For mining area directly affiliated to the Sanshandao Gold Mine, underground water supplementing the pit may come from seawater, Quaternary pore water and bedrock brine in different chemical environments. Ions in the water body undergo changes and exchanges under varying physical and chemical environments and thus it presents different hydrochemical characteristics. Main occurrence of granite in the mining area is affected by different hydrochemical environments, and its physical and mechanical properties and deformation characteristics have witnessed changes to varying degrees. Studying hydrochemical composition and content of underground water in mining areas can clarify degree of physical and chemical damage to the mechanical properties of rocks and provide a scientific basis for design of mining and development projects; moreover, selecting a certain index that can reflect environmental state of underground water and using the isotope and hydrochemical synthetic analysis method also helps determine source of replenishment of a certain water gushing point and predict its evolution direction. So analysis of hydrochemical composition of underground water in mining area has important research significance.

To study composition and content of chemical composition of underground water in the mining area, 15 representative water points are selected from the depth ranging from -510 m to -725 m for water samples for analysis and testing on hydrochemical components. The sampling locations are listed in Table 5-2.

Sampling statistics of analysis on hydrochemical composition of underground water in the mining area directly under Sanshandao Gold Mine   Table 5-2

Serial number	Coding of water sample	Sampling site	Nature of water penetration
1	510-1	Stope 516	Drill hole running water on the left side of the roof
2	510-2	North Lane F3 fault in the middle section at -510 m	Running water from the roof and right side
3	525-1	Horizontal sublevel roadway at -525 m	Roof water seepage
4	525-2	Panel 552 No. 7 Mining Union	Water seepage at the left side of the roadway
5	555-1	Near the return air shaft of North Lane Line 1720	Flowing water at crack on the roof
6	555-2	Grand Lane F3 fault in the middle section	Gushing water from the roof and right side
7	555-3	Stope S12186 NO. 10 Mining Union	Running water from crack on the right

5 Rock Chemical Damage Induced by Deep Hydrochemical Environment

continued

Serial number	Coding of water sample	Sampling site	Nature of water penetration
8	600-1	Ramp entrance in the middle section at −600 m	Water seepage from the tunnel roof and left side
9	600-2	Stope S13165	Water seepage from the crack on the roof
10	600-3	North Lane F3 fault in the middle section at −600 m	Gushing water from the roof and two sides
11	645-1	Near the Line 1480 of the North Lane in the middle section at −655 m	Drill hole running water from the right side
12	645-2	Stope S14156 Mining Alley	Water seepage from the left side of the roadway
13	690-1	Ramp entrance in the middle section at −690 m	Drill hole running water from the left side of the roadway
14	690-2	Near the Line 1460 of South lane in the middle section at −690 m	Roof water seepage
15	725-1	Horizontal reverse lane at ramp at −725 m	Water seepage from the right side of the roadway

Results of hydrochemical analysis of each water sample are listed in Table 5-3. By analyzing data in Table 5-3, it can be found that the main anion components of deep underground water in the mining area are Cl^-, SO_4^{2-} and HCO_3^-, and main cations are Na^+, Ca^{2+} and Mg^{2+}. Concentration of Cl^- is 11728.50-25382.65 mg/L and the average concentration is 18737.90 mg/L; concentration of SO_4^{2-} is 1325.37-3089.85 mg/L and the average concentration is 2206.01 mg/L; concentration of HCO_3^- is 30.80-389.54 mg/L and the average concentration is 179.56 mg/L; concentration of Na^+ is 6450.37-12329.65 mg/L and the average concentration is 9067.44 mg/L; concentration of Mg^{2+} is 475.51-1898.21 mg/L and the average concentration is 962.08 mg/L; concentration of Ca^{2+} is 581.44-1841.09 mg/L and the average concentration is 1089.31 mg/L; total mineralization degree is 21255.45-44015.64 mg/L and the average value is 32320.70 mg/L; the pH value is 4.68-7.14, and the average pH value is 6.18.

Sampling statistics of analysis on hydrochemical composition of underground
water in the mining area directly under Sanshandao Gold Mine   Table 5-3

Coding of water sample	Concentration of main anion(mg/L)			Concentration of main cation(mg/L)			Total salinity (mg/L)	pH value
	Cl^-	SO_4^{2-}	HCO_3^-	Na^+	Mg^{2+}	Ca^{2+}		
510-1	16523.88	2261.24	297.12	8847.47	487.02	1045.30	29484.26	6.43
510-2	16025.40	1807.01	100.94	6955.79	1124.45	965.45	26951.54	6.20
525-1	13444.68	2057.21	234.41	7497.21	731.36	898.88	24901.21	7.05
525-2	19365.27	2342.32	79.50	10054.26	668.48	854.01	33423.41	5.89
555-1	20894.24	2145.96	175.90	9784.20	475.51	1164.71	34844.40	5.22
555-2	15998.05	2167.64	141.46	7751.26	746.70	1224.83	28111.65	6.38
555-3	11728.50	1325.37	268.11	6450.37	733.50	746.10	21255.45	7.14
600-1	21828.79	2365.68	231.56	10789.73	982.22	1432.36	37841.52	6.11
600-2	19154.18	2298.28	231.70	9452.80	754.51	581.44	32512.91	5.71
600-3	21149.15	2356.63	54.55	10052.42	1384.30	1841.09	36895.20	6.80
645-1	19422.77	3089.85	389.54	6894.24	912.45	1288.27	32125.11	4.68
645-2	25382.65	2877.16	255.53	12329.65	1898.21	1154.32	44015.64	6.27
690-1	21583.90	2132.05	103.36	9446.21	1463.83	1397.62	36182.24	5.97
690-2	22114.57	2415.61	98.85	11336.05	1093.48	888.97	38110.12	6.04
725-1	16452.45	1448.19	30.80	8369.97	975.24	856.32	28155.89	6.75

The most abundant anion and cation in deep underground water samples in the mining area is respectively Cl^- and Na^+. It is mainly Cl-Na water. Test results on salinity reveal that it is characterized by highly mineralized salt water. By taking into account dilution effect of filling water in production operation of the mining area, highly mineralized bedrock brine exists in replenishing water source of the sampling water gushing point. As depth of the sampling increases, salinity of the water sample tends to increase. The pH value of water sample taken from the gushing point at -645 m horizontal borehole is the minimum at 4.68; pH value of the water sample taken from S12186 quarry No. 10 sampling unit at -555 m is the maximum at 7.14, and pH value of the remaining water samples most stays between 6 and 7, and most are faintly acidic. In experimental research on chemical damage of granite, it has been found that the acidic NaCl solution has a certain chemical corrosion effect on granite and main ion components of underground water samples collected from the deep part of the mining area also contain Cl^- and Na^+, and those water samples are mostly weak acidic.

It can be predicted that due to existence of a specific hydrochemical environment, underground water in the mining area directly under Sanshan Island may cause chemical damage on granite in the deep and have a certain impact on its physical and mechanical properties. Designing and constructing of deep mining and excavation projects in mining areas should factor in the above-mentioned impacts that might be imposed on rocks by hydrochemical damage in order to ensure safe and efficient development of deep mining projects.

5.1.3 Salinity test of underground water

Salinity of seawater is quantitative measurement of salinity in seawater. It is one of the most important physicochemical properties of seawater and is closely related with coastal runoff, precipitation and sea surface evaporation. Distribution changes of salinity also compose a major factor affecting and constraining distribution and changes of other hydrological elements. Monitoring temporal and spatial trend of salinity changes at various points can provide basis for judging connectivity between bedrock brine and the upper seawater.

By collecting water samples from original points at different levels and following requirements of marine monitoring regulations (Marine Monitoring, 1999), this paper uses induction salinometer (Figure 5-4), refers to analysis on conductivity ratio analysis of standard seawater with a salinity of 35 at 28°C, and conducts measurement of salinity in seawater in the Key Laboratory of Marine Chemistry in Ocean University of China. Detailed data are seen in Table 5-4.

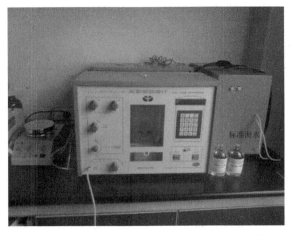

Figure 5-4 Induction salinity meter for measuring salinity of ancient seawater and corresponding standard seawater

Record of analysis on seawater power-induction salinity Ocean Monitoring and Testing Center of Ocean University of China (Q/HDJC 3176C—2007)   Table 5-4

Serial number	Time	Sample depth(m)	Coding	Measured value			Notes
				S_1	S_2	S (Mean value)	
1	7.3	375	SZ(S)140710-001	33.154	33.154	33.154	
2	7.5	600	SZ(S)140710-002	32.385	32.385	32.385	
3	7.5	645	SZ(S)140710-003	35.379	35.378	35.379	
4	7.5	690	SZ(S)140710-004	42.878	42.877	42.878	
5	7.5	735	SZ(S)140710-005	38.464	38.463	38.464	
6	7.5	760	SZ(S)140710-006	19.968	19.968	19.968	
7	7.6	375	SZ(S)140710-007	34.317	34.317	34.317	
8	7.6	600	SZ(S)140710-008	31.119	31.118	31.119	
9	7.6	645	SZ(S)140710-009	45.616	45.617	45.617	Dilution test
10	7.6	670	SZ(S)140710-010	40.863	40.863	40.863	
11	7.6	735	SZ(S)140710-011	45.580	45.581	45.581	Dilution test
12	7.6	765	SZ(S)140710-012	38.479	38.479	38.479	
13	7.9	330	SZ(S)140710-013	32.964	32.963	32.964	
14	7.9	600	SZ(S)140710-014	31.278	31.278	31.278	
15	7.9	645	SZ(S)140710-015	34.856	34.857	34.857	
16	7.9	690	SZ(S)140710-016	40.917	40.916	40.917	
17	7.9	690	SZ(S)140710-017	42.813	42.813	42.813	
18	7.9	780	SZ(S)140710-018	54.129	54.130	54.130	Dilution test

5.1.4 Seepage field simulation analysis in deep mining area

Sanshandao Gold Mine, a rare large-scale coastal metal mine in China, has scanty successful experience in safety production. Compared with onshore mining, mining of seabed minerals is up against more severe hydrogeological problems in mines. Under combined action of seawater pressure and seepage field inside the rock, movement and deformation rules of the rock and distribution characteristics of the seepage field caused by deep mining disturbance are more complicated. Crack expansion and activation of faults may bring potential dangers such as water inrush, water inrush and even sea water intrusion to coastal mining. On the ground of field investigation of hydrogeological environment in mines and indoor rock mechanics tests, this paper applies the finite difference method numerical software FLAC3D to simulate and calculate distribution characteristics of the seepage field of rock in the future deep mining process and to provide certain guidance for design and construction of deep mining project, and mine flood prevention and control

(1) Geometric model and physical and mechanical parameters of rocks

Based on future deep mining schedule and occurrence of ore bodies in the mining area directly under Sanshandao Gold Mine, a three-dimensional numerical calculation model is established by use of FLAC3D software. In the coordinate system of the model, proneness direction of the ore body is the X-axis, thickness direction of the ore body is the Y-axis, and the vertical direction is the Z-axis; X, Y, and Z directions of the model are respectively 1100 m, 750 m and 950 m in length and its elevation ranges from -335 m to -1305 m; the conceptual model based on engineering geological conditions of the deposit can be roughly divided into five zones covering hanging wall rock body group, ore body group, footwall rock body group, fault group and filling body group.

Physical and mechanical parameters of main rocks in the deep mining area are selected mainly based on results of the indoor rock mechanics test. When factoring into the scale effect, parameters obtained from the test can be appropriately reduced. Mechanical parameters of the rock and backfill are shown in Table 5-5.

As rocks in deep Sanshandao Gold Mine are subject to the impact of underground water with high salinity and complex hydrochemical composition in the area for a long time. Computation of data calls for taking into account the effect of hydrochemical damage on the physical and mechanical properties of rock. The aforesaid sampling analysis on hydrochemical composition in the mining area reveals that the minimum pH value in the test water sample is 4.68. The numerical calculation presumes that deep underground hydrochemical environment of the Sanshan Island mining area is 4 and takes into account hydrochemical damage effect of the rock. Although this method is conservative, it improves safety coefficient of the numerical calculation results and thus has a positive side in practice. So, on the ground of the model calculation parameters in Table 5-5, calculation parameters are appropriately adjusted when factoring in hydrochemical damage effect of the rock. Rock parameters after adjustment are listed in

Table 5-6.

Physical and mechanical parameters for data simulation
(without considering hydrochemical damage effect of rock)   Table 5-5

Site	Concentration (kg/m³)	Bulk modulus (GPa)	Shear modulus (GPa)	Poisson's ratio	Cohesion (MPa)	Internal friction angle(°)	Compresion strength (MPa)	Tensile strength (MPa)	Osmotic coefficient (m/s)
Hanging wall	2706	3.37	2.8	0.21	11.4	31.3	48.2	4.37	9.13×10^{-7}
Footwall	2635	5.48	3.45	0.24	42.8	36.9	60.1	5.98	6.72×10^{-7}
Fault	1925	0.595	0.378	0.26	0.128	18.4	20.6	0.0198	1.38×10^{-5}
Ore body	2710	2.51	1.35	0.19	21.5	33.1	57.6	3.44	8.32×10^{-7}
Filling border	1610	0.304	0.0245	0.17	0.261	35.6	7.43	0.212	4.78×10^{-6}

Physical and mechanical parameters of numerical simulation after
considering hydrochemical damage effect of rock   Table 5-6

Site	Concentration (kg/m³)	Bulk modulus (GPa)	Shear modulus (GPa)	Poisson's ratio	Cohesion (MPa)	Internal friction angle(°)	Compresion strength (MPa)	Tensile strength (MPa)	Osmotic coefficient (m/s)
Hanging wall	2665	3.28	2.58	0.22	8.9	28.0	36.9	3.77	1.09×10^{-6}
Footwall	2595	5.33	3.18	0.26	33.6	33.0	46.0	5.16	7.99×10^{-7}
Fault	1707	0.521	0.313	0.31	0.0903	14.8	14.2	0.0154	1.81×10^{-5}
Ore body	2669	2.44	1.24	0.20	16.9	29.6	44.1	2.97	9.89×10^{-7}
Filling border	1507	0.281	0.0214	0.19	0.194	30.2	5.41	0.174	5.97×10^{-6}

(2) Initial stress field and boundary conditions

Presume that mechanical boundary conditions of the model with the bottom ($z=0$) as the fixed constrained boundary, the top ($z=950$) as the stress boundary and the rest boundaries of the model as unidirectional; the seepage field calculation boundary is the seepage free boundary. Stress boundary of the model is applied based on results of field crustal stress measured in field test in the mining area directly under Sanshandao Gold Mine. Changing rules of maximum horizontal principal stress, minimum horizontal principal stress, and vertical principal stress with changes in depth as seen in equation (5-1) -equation (5-3):

$$\sigma_{h, max} = 1.433 + 0.043H \quad (5-1)$$
$$\sigma_{h, min} = 0.536 + 0.024H \quad (5-2)$$
$$\sigma_v = 0.838 + 0.027H \quad (5-3)$$

where $\sigma_{h, max}$ —maximum horizontal principal stress, MPa;

$\sigma_{h,\ min}$ —minimum horizontal principal stress, MPa;

σ_v —vertical principal stress, MPa;

H —buried depth, m.

(3) Designing scheme for simulation schemes

In order to better simulate deep mining process of Sanshandao Gold Mine, a phase-based excavation method is used to simulate deep mine excavation. During deep development process of the mine, a middle section of production at every 45 m for vertical height is set up. After ore bodies in a specific middle section are almost mined, the next middle section of the ore body is mined. So in numerical calculation, the plan of excavation every 45 m is adopted and excavation is carried out in 10 phases from the elevation at −450 m to that at −1000 m. Only ore bodies 42 m in height are mined in each middle section and the 3 m pillar left on top is not mined to ensure safety of the upper adjacent stope floor. Schematic diagram of the overall model, schematic diagram of the excavation model, the initial stress field and the initial pore water pressure are shown in Figure 5-5 to Figure 5-8.

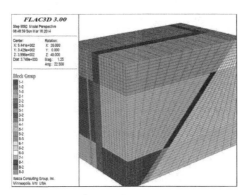

Figure 5-5 Schematic diagram of the overall model

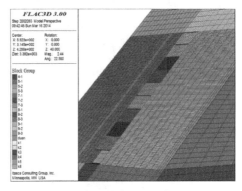

Figure 5-6 Schematic diagram of the excavation model

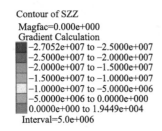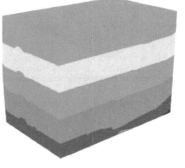

Figure 5-7 Initial stress field

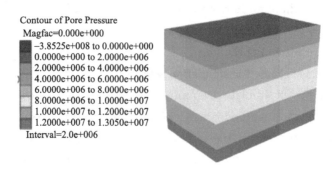

Figure 5-8   Initial pore water pressure

Based on results of flow-through coupling analysis, this paper extracts the overall distribution of pore water pressure from the excavation beginning to phase 10, as is shown in Figure 5-9.

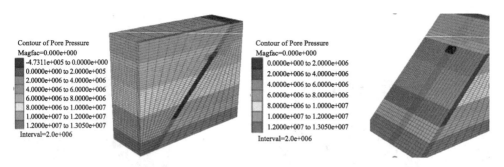

(a) Nephogram of pore water pressure distribution in the first phase of excavation

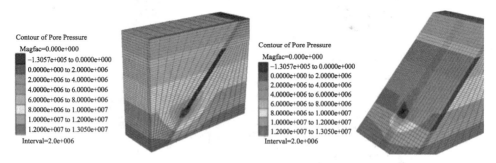

(b) Nephogram of pore water pressure distribution in the tenth phase of excavation

Figure 5-9   The overall model seepage field

Simulation calculation results show that changing range of pore water pressure in the mining area is mainly concentrated in the excavation area of the ore body. With continuous mining of the ore body, permeability coefficient of the filling body is relatively large and a good permeability channel is formed along the excavation body. Area along direction of the ore body is affected by water-conducting zone of the F3 fault. Pore water pressure near the fault has high gradient in changes of seepage pressure, resulting in high permeability pressure. With excavation and unloading effect of rocks in the deep high-stress area, rock cracks continue to develop and the permeability witnesses changes, and after the filling body replaces the original ore body, its permeability also witnesses changes, which will expand the influencing depth and scope of coupling effect of seepage stress. When the simulated mining reaches the level of -1050 m, corresponding relationship between seepage field distribution and the position of the middle section of the entire mining area is seen in Figure 5-10.

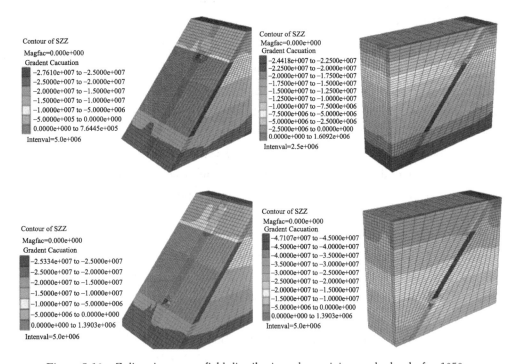

Figure 5-10　Z-direction stress field distribution when mining at the level of -1050 m

Due to mining unloading effect and continuous erosion of underground water on surrounding rocks, rock cracks continue to develop. With continuous mining of ore bodies, range of changes in the seepage field of rock in the stope area continues to expand. At phase ten of simulation mining, range of change in the seepage field has been extended to F3 fault. From this analysis, it can be seen that in the future mining process of Sanshandao Gold Mine below a depth of one kilometer, seepage will have an impact on stability of the deep stope. If surrounding rock of the stope forms a new water channel under the action of excavation unloading, high permeability water pressure and underground water, the coupling effect of seepage and stress will be more obvious, which will have a greater impact on stability of the stope.

5.2 Damage mechanism and time-dependent characteristics of granite under the action of acidic chemical solutions

5.2.1 Granite deformation and failure test under chemistry-seepage coupling effect

Rock is an aggregate of natural geological products formed by cementation and bonding of particles or crystals. Pore structures such as micro-cracks, mineral joints, intergranular voids and lattice defects are found inside. Its physical and mechanical properties mainly depend on connection between rock mineral composition and particles as well as microcracks that exist in its interior. Affected by environmental factors, rock materials in the project will inevitably be subject to water and various chemical erosion. Interaction between water and rock weakens connection of mineral particles, corrodes mineral particles or crystals and changes microstructure and physical state of rocks, thereby causing continuous degradation of rock mechanical properties. Change of microstructure leads to change of its macro-mechanical properties. The fundamental reason is that parameters of mechanical properties of rocks witness changes to varying degrees. Based on erstwhile studies and by carrying out uniaxial test, triaxial compression tests and splitting strength tests, deformation and strength characteristics of granite under varying water velocity environments and chemical solution erosion environments are further explored for the purpose of offering basis to explore mechanism of failure of chemistry-seepage coupling effect before setting up the constitutive model of the granite failure process under hydrochemical environment.

5.2.2 Experimental procedures

(1) Preparation of rock specimen

Specimens used in the experiment are taken from granite of Sanshandao Gold

Mine of Shandong Gold Mining Co., Ltd. The sampling point is in the north lane of the transportation main lane at −600 m in the mining area directly under the Sanshandao Gold Mine. During field sampling, the north lane at −600 m was in the process of excavation from south to north and the blasting operation produced massive granite rocks. To facilitate processing of the specimens, a cubic, regular rock is selected as the sample. Moreover, some cylindrical core processing specimens obtained from field drilling rig prospecting operations are also selected. After the rock sample is retrieved from the site, the part with rock surface 3-5 cm in thickness is removed since rock samples are affected by certain weathering during the transportation and storage process. It helps ensure freshness of the rock sample. Use a drill to drill a cylindrical rock sample with a diameter of 50 mm and then cut the rock into samples 100 mm and 25 mm in height on a saw stone cutting machine. Apply a stone grinder to smooth two sections of the rock sample to form a $\phi 50$ mm×100 mm column, $\phi 50$ mm×25 mm disc-shaped specimen. See Figure 5-11 for details. The processing accuracy of each specimen including parallelism, straightness and perpendicularity is controlled within the specified range of *Regulations in Rock Test for Water Conservancy and Hydropower Project* SL/T 264—2020. After all specimens are prepared, keep the specimens in a ventilated and dry place for more than half a month and clean surface of the specimens for later use. Through thin slice identification on composition of granite specimens, the specimens are mainly sericitized granite and phenocryst granite.

 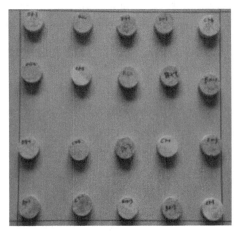

Figure 5-11　Granite specimens used in the experiment

(2) Chemical solution configuration

Chemical composition of the actual underground bedrock brine is very complicated. Ions and complexes in the aqueous solution exist in various forms, and their ion concentration and pH value are constantly changing with time and space. Main ions contained in most underground water are Ca^{2+}, Mg^{2+}, Na^+, K^+, Cl^-, SO_4^{2-}, HCO_3^- and so on. Under the condition of indoors experiments, the reaction process of rocks under the action of underground bedrock brine in nature is directly studied. Affected by complex chemical composition and multiple uncertain factors, the research process is complicated and is difficult to obtain ideal test results under limitation of time and conditions. Related studies have shown that chemical reaction between rock and the chemical solution is largely controlled by pH value of the solution. Sanshandao Gold Mine, where the rock specimens are sampled, is a coastal mine. Supplementing underground water in the mining area may hail from sea water, Quaternary water and bedrock brine, so it has high concentration of Cl^- ions and a slightly acidic pH value. Mineral components of deep granite in the mining area contain a certain amount of feldspar, and feldspar minerals have significant corrosion and dissolution effects in acidic environments. So in order to observe failure effect of the hydrochemical solution on the rock in a short time, and by combining with main anion and cation components in the actual underground water of Sanshandao Gold Mine and complying with the principles of appropriately increasing of concentration of reactants, shortening of reaction time and speedy testing, the following chemical solutions are configured in Table 5-7.

Instruments for configuring chemical solution in the experiment is seen in Figure 5-12. Preparation and calibration of chemical solution in the test are carried out as below:

1) Take a certain amount of concentrated hydrochloric acid with a graduated cylinder, pour it into a 500 mL volumetric flask, add distilled water to dilute it to 500 mL, cover it with a glass stopper, and shake it evenly.

2) Apply an electronic scale to accurately weigh 3 pieces of anhydrous sodium carbonate, the standard substance, weighing 0.11-0.14 g, put them in a 250 mL beaker, add distilled water of 20-30mL for dissolution, and add 2 drops of methyl orange indicator to each piece.

3) Drop HCl to the solution until it turns from yellow to orange, which marks an end point. Calculate concentration of the HCl standard solution based

on quality of the solid sodium carbonate and volume of HCl solution consumed during titration.

4) Add the prepared HCl standard solution into the acid tank and dilute it with distilled water in the acid tank to concentration required in chemical solution in Table 5-7 based on concentration of the HCl standard solution. While diluting, weigh an appropriate amount of NaCl powder with an electronic scale and add it to the acid tank for adequate stirring.

Chemical solutions used in the test   Table 5-7

Elements of solution	Concentration(mol/L)	pH value
NaCl	0.01	2
NaCl	0.01	4
NaCl	0.01	7
Distilled water	—	7

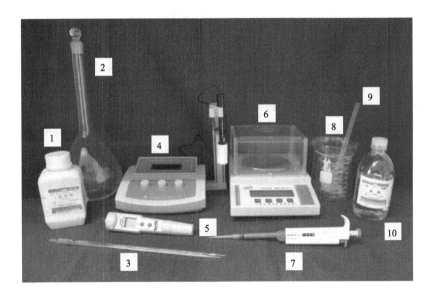

Figure 5-12 Equipment for configuring chemical solutions
(Note: 1—solid NaCl; 2—volume flask; 3—pipette etched tube; 4—pH analyzer; 5—pH test pen;
6—analytical scale; 7—pipette gun; 8—beakers of different specifications; 9—glass rod;
10—concentrated hydrochloric acid)

(3) Immersion test device that realizes chemistry and percolation

Due to widespread existence of water chemical solutions in nature, rocks in underground projects are mainly affected by the action of underground water fluids. Environmental conditions and characteristics of underground rocks in the state of rich water are mainly seen below:

1) Affected by the action of underground hydrodynamic field (underground rock excavation, high water pressure and so on), underground water is generally in a flowing state and its flow often constantly changes with time.

2) Flow of underground water not just erodes and dissolves rocks, ions in water that can react with certain components in the rock can affect mechanics or hydraulic nature of rock by taking effect on and transforming composition of rock and the fracture surface through chemical corrosion.

3) Hydraulic pressure of underground water changes crack surface and force state among particles in the rock, accelerates expansion of rock cracks, reduces effective stress in the rock or brings down strength of the rock.

It can be thus seen that underground water has three main effects on rocks namely physical effects (lubrication, softening and siltification, and strengthening of bound water), chemical effects (dissolution and hydrolysis, ion exchange, redox reaction), mechanical action (effect of pore hydrostatic pressure and hydrodynamic pressure). Study on the interaction between water and rock must consider coupling of stress field, chemical field and seepage field and study the response mechanism of comprehensive physical, chemical and mechanical properties of rocks in a stress-chemical-seepage coupling test environment.

Specific to the aforesaid problems, a set of test equipment that simulates and reproduces chemical dissolution and seepage erosion effect of flowing water on rocks in the natural environment is developed and assembled, as shown in Figure 5-13. Immerse the granite specimen in a pre-configured chemical solution for a period of time, which approximates the chemical reaction process between granite and chemical solution; with dynamic impact of the water pump, the chemical solution flows continuously in the test container to form a certain seepage water pressure, which simulates the seepage and flow effect of chemical solutions in granite specimens; after a period of immersion test, the specimens are processed and tested on a rock mechanics testing machine. Test results are applied to analyze deformation characteristics, deterioration degree of physical and mechanical parameters and so on after erosion of hydrochemical solution so as to achieve the purpose of studying damage effect of graniteunder stress-chemical-seepage multi-field coupling.

Test devices in Figure 5-13 consist of an acid tank container a booster self-priming pump and a pipe. The acid tank is made of polyvinyl chloride synthetic material and can withstand strong acids and alkalis. The size is 500 mm × 350 mm × 350 mm and it is 15 mm in thickness. Main body of the acid tank adopts a double-layer design with inner and outer tanks. The bottom and top of

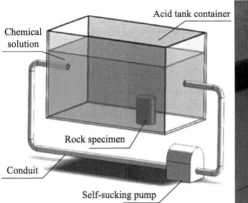

Figure 5-13 Schematic diagram and photo of immersion
experiment devices for chemical seepage coupling

the outer tank are equipped with special adhesive seals. No leakage will occur after use on a long-term basis and thus safety of the test can be ensured. The booster self-priming pump has high suction stroke and high efficiency. Pressure of the water pump can be adjusted by setup of power of water pump. It can realize automatic circulation of chemical solution in the acid tank and meet requirements of the test conditions. Self-priming pumps and conduits also meet test conditions for resistance against acid and alkali corrosion.

Before the immersion test, each specimen is numbered in groups. Specimens are then placed in an acid tank containing a pre-prepared chemical solution based on groups following such indexes as size, quality, porosity and so on. After preparation for the test is completed, switch on power of the water pump, and then water flow is drawn from water near bottom of the acid tank that then returns to the container through conduit after flowing through the water pump. It thus achieves repeated circulation flow. Based on research purpose and flow rate of underground water gushing at the level of -600 m in Sanshandao Gold Mine, test environments with three flowing speeds are set up by adjustment of working power of the pump, which are respectively static water state ($v=0$ mm/s), low circulating water flow rate state ($v=200$ mm/s) and high circulating water flow rate state ($v=400$ mm/s). As underground water is often excessive in local environments, and for the purpose of realizing water circulation of chemical solution, the corresponding chemical immersion solution of each specimen is set at 2 L. The test is carried out at room temperature, and the temperature does not change much during the immersion process, so impacts of temperature on mechanical properties and chemical reactions of the rock are overlooked. During the immersion test, quality, elastic wave velocity and other physical and mechanical indicators of all rock specimens are measured at regular intervals and samples are taken to analyze ion composition and pH value of the chemical solution. After immersing for 60 days, specimens are taken out and corresponding mechanical tests are carried out on the rock mechanics testing machine.

5.2.3 Uniaxial compression test

(1) Test plan

This test applies the TAW-2000 computer-controlled electro-hydraulic servo rock testing machine that can also be used to study mechanical properties and shear characteristics of rocks in a variety of environments and can automatically complete uniaxial compression test of the rock. The testing machine has a limit control function, and can enter automatic protection when axial deformation, radial deformation time and other parameters reach the limit or preset value, the sample is broken, the oil circuit is blocked and the oil temperature is too high. Rock specimens in the test are prepared following the specification of cylinder size of $\phi 50$ mm $\times$ 100 mm and then are immersed in a chemical solution. After being sealed and immersed in a static water environment for 60 days, they are processed and then subject to a mechanical test on the testing machine.

(2) Analysis on experimental results

The stress-strain curve of uniaxial compression test on granite specimen is seen in Figure 5-14. By combining with the typical full stress-strain curve diagram of rock uniaxial compression deformation in Figure 5-15, characteristics and rules of uniaxial compression deformation of granite before and after action of chemical solution are analyzed in detail as below:

1) The oa section is the initial crack compaction phase. After action of the hydrochemical solution, the original openstructural planes or micro-cracks of some specimens gradually close, rock is compacted to form early-stage nonlinear deformation, and σ-ε curve shows an obviously concave shape. NaCl with pH=2 has the most significant effect. $d\sigma/d\varepsilon$ increases with expansion of stress, $d^2\sigma/d\varepsilon^2 > 0$. It reveals that chemical solution erodes and dissolves the rock, which causes the pores to increase or become longer and the initial crack compaction phase to extend. Deformation at this phase is more obvious for rocks with developed fractures but is less so for hard and dense rocks.

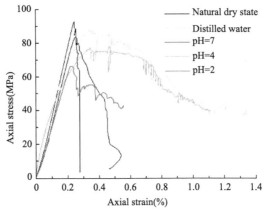
(a) Natural dry state and after action of different hydrochemical solutions (v=0 mm/s)

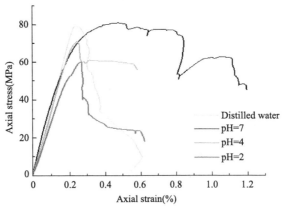
(b) After action of different hydrochemical solutions (v=200 mm/s)

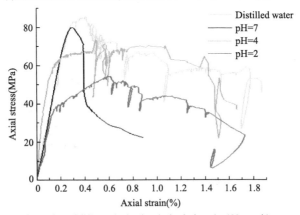
(c) After action of different hydrochemical solutions (v=400 mm/s)

Figure 5-14　The uniaxial compressive stress-strain curves of granite specimens during natural drying and after action of different hydrochemical solutions

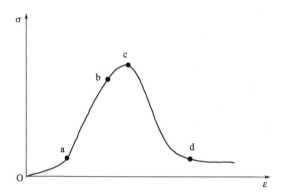

Figure 5-15 Stress-strain curve of rock uniaxial compression test

From Figure 5-14, it can be found that the compaction phase of granite specimens under natural dry conditions is short and the entire compaction process is transient. It shows that the specimens are relatively dense and the deformation enters the elastic deformation phase earlier; after being subjected to hydrochemical action, compaction phase of the granite specimens becomes more obvious and most enter the elastic deformation phase after a relatively long concave phase. Length of the pore fracture compaction phase is closely related to degree of fracture development in the specimen. Greater erosion and dissolution of the granite specimen by chemical solution leads to more significant increase of pores and longer compaction phase. It should be noted that due to heterogeneity of the granite specimens and selectivity of chemical erosion, some specimens with insignificant compaction phase are also found.

2) The ab section is the phase marked by stable development from elastic deformation to micro-elastic fracture. In this phase, $d\sigma/d\varepsilon = k$ (k is a constant, lasticity modulus), $d^2\sigma/d\varepsilon^2 = 0$ and σ-ε curve is approximately a straight line. k also reflects slope of the straight line. After action of the hydrochemical solution, slope of the curve of the granite specimens at this phase decreased to varying degrees, indicating that granite is softened by chemical solution and elastic modulus decreases.

3) The bc section is the development phase marked by unstable fracture. At the phase of occurrence, development, and penetration of micro-cracks or at the phase marked by stable development of micro-fracture, $d\sigma/d\varepsilon$ decreases with increase of stress, $d^2\sigma/d\varepsilon^2 < 0$. Stress-strain curve of the granite specimen in a natural dry state is basically a straight line before peak, and no obvious yield phase

is traced, but a relatively obvious peak point is shown. After action of the hydrochemical solution, stress-strain curve of some granite specimens begins to bend downward and shows evident yield characteristics when stress reaches a certain value; peak strength decreases with decrease of pH value of the solution and increase of flow rate of circulating water. Peak strength of uniaxial compression decreases the most when compared with that under natural dry state, reaching 41.38% after acting in NaCl solution with flow rate $v=400$ mm/s and pH=2 for 60 days. This phase is related to changes in the microstructure before and after erosion on rocks by chemical solution and to softening degree of rocks.

4) The cd section is the post-fracture phase. Failure of granite in a natural dry state has obvious brittle characteristics. Specimens go through abrupt failure during the loading process and are accompanied by a loud cracking sound. Stress-strain curve of granite in a natural dry state reveals that after reaching the peak, the curve rapidly develops downwards and shows a relatively large stress drop; the curve undergoes a short step-like drop after reaching the peak with stress quickly plummeting to near 0. The failure means is unstable brittle failure. After action of the hydrochemical solution, rock has softened and its brittleness during failure is weakened. The specimen still shows large deformation development after reaching peak point, indicating that granite has a tendency to convert from brittleness to ductility under the action of the chemical solution.

Moreover, due to difference in natural properties of the test granite, heterogeneity of the specimens and selectivity of chemical erosion, microstructure changes and degree of corrosion softening of the rock slightly differ after chemical solution erosion, which results in varying mechanical characteristics after action of chemical solutions on some specimens.

(3) Impact of chemical solution on the E and μ values of granite

Elastic modulus and Poisson's ratio are one of the important parameters that characterize rock deformation characteristics. Due to heterogeneity of rock, stress-strain relationship of rock specimen is a curve rather than a straight line. Two generally-accepted methods are used for determining elastic modulus and Poisson's ratio of rocks:

1) Secant elastic modulus and Poisson's ratio: slope of the connecting line between the point on the rock stress-axial strain curve of a specific ratio (usually 50%) corresponding to compressive strength. The calculation equation is as follows:

$$E = \sigma_{c(50\%)} / \varepsilon_{h(50\%)}$$
$$\mu = \varepsilon_{d(50\%)} / \varepsilon_{h(50\%)}$$
(5-4)

where $\sigma_{c(50\%)}$ —50% of the compressive strength of the specimen;

$\varepsilon_{h(50\%)}$ —the corresponding axial compressive strain at $\sigma_{c(50\%)}$;

$\varepsilon_{d(50\%)}$ —the corresponding radial tensile strain at $\sigma_{c(50\%)}$.

2) Average elastic modulus and Poisson's ratio: it is obtained by calculating mean slope of the approximate straight line on the axial stress-strain curve. The calculation equation is as follows:

$$E = \frac{\sigma_b - \sigma_a}{\varepsilon_{hb} - \varepsilon_{ha}}$$
$$\mu = \frac{\varepsilon_{db} - \varepsilon_{da}}{\varepsilon_{hb} - \varepsilon_{ha}}$$
(5-5)

where σ_a —stress value at a, starting point of the straight line on the curve of relationship between stress and longitudinal strain;

σ_b —stress value at b, ending point of the straight line on the curve of relationship between stress and longitudinal strain;

ε_{ha} —axial strain of a, starting point of the straight line;

ε_{da} —radial strain of a, starting point of the straight line;

ε_{hb} —axial strain of b, ending point of the straight line;

ε_{db} —radial strain of b, ending point of the straight line.

Magnitude of average elastic modulus depends on slope of the approximately-straight part of the stress-strain curve, and is less affected by test conditions, so it can more accurately reflect impacts of hydrochemical solution on deformation properties of granite. Following equation (5-5), the average elastic modulus of each granite specimen is calculated and the result is seen in Figure 5-16.

It can be learned from Figure 5-16 that after action of different chemical solutions, deformation of the granite specimen under the uniaxial compression test conditions increases and elastic modulus decreases to varying degrees. After the granite specimen is immersed in NaCl solution with a pH=2 and a flow rate $v=$ 400 mm/s for 60 days, decrease amplitude of the elastic modulus reaches the maximum at 27.98%. The elastic modulus of granite is most sensitive to pH value of the chemical solution; when composition of the chemical solution is the same, higher circulating water flow rate means more significant decrease in elastic modulus of granite; although heterogeneity of the rock and difference in chemical corrosion leads discrete strength and deformation of some specimens,

5 Rock Chemical Damage Induced by Deep Hydrochemical Environment

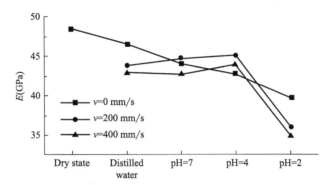

Figure 5-16 Impact of chemical solution on elastic modulus of granite

change trend of elastic modulus of granite after action of the chemical solution basically aligns with the aforesaid rules.

Following equation (5-5), Poisson's ratio before and after immersion of the granite specimen is calculated and the result is seen in Figure 5-17. Poisson's ratio of granite is also sensitive to hydrochemical environment; different from elastic modulus, Poisson's ratio of the granite specimens increased to varying degrees after chemical seepage. After being immersed in NaCl solution with pH=2 and a flow rate of 400 mm/s for 60 days, Poisson's ratio of the granite specimens increased by 32.25% when compared to that under natural dry conditions.

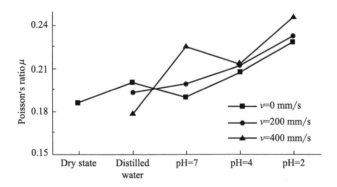

Figure 5-17 Impact of chemical solution on Poisson's ratio of granite

5.2.4 Triaxial compression test

(1) Test plan

The rock specimen in the test is prepared based on a cylinder size of $\phi 50$ mm×100 mm and then is immersed in a chemical solution. It is sealed and immersed in static water for 60 days before being subject to a mechanical test. The test is done at room temperature and temperature does not change much during the immersion process. Impact of temperature on mechanical properties and chemical reactions of the rock is not considered.

Based on field stress measurement results of Sanshandao Gold Mine, the maximum principal stress at the level of -600 m in the mining area is about 29.53 MPa. Confining pressures of specimens immersed in each solution are respectively 5 MPa, 10 MPa, 20 MPa and 30 MPa. TYS-500 triaxial testing machine is used in this test. It applies load-controlled loading mode and the loading rate is 1 MPa/s.

(2) Stress-strain curve of triaxial compression test

Based on the experiment, the triaxial compression stress-strain curves of granite specimens during natural drying and with action of four different water chemical solutions are obtained, as are shown in Figure 5-18.

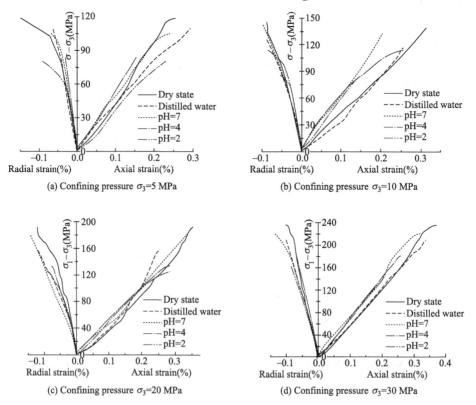

Figure 5-18 Triaxial compressive stress-strain curves of granite specimens after natural drying and action of different hydrochemical solutions

5 Rock Chemical Damage Induced by Deep Hydrochemical Environment

Test results show that when compared with natural drying, granite specimens with the action of different water chemical solutions show different degrees of softening. It can be seen from Figure 5-18 that the triaxial peak compressive strength of granite increases significantly with rise of confining pressure and impact of hydrochemical solution narrow down the increase. pH value of the chemical solution is a controlling factor that affects triaxial peak compressive strength of the specimen. After granite is immersed in a 0.01 mol/L NaCl solution with pH=2 for 60 days, its peak strength shows maximum range of decrease, and as it decreases respectively by 31.4% ($\sigma_3 = 5$ MPa), 39.9% ($\sigma_3 = 10$ MPa), 31.2% ($\sigma_3 = 20$ MPa) and 27.8% ($\sigma_3 = 30$ MPa) when compared to the natural drying. Measured from deformation characteristics of granite specimen, the compaction process of the specimen is prolonged with action of hydrochemical solution, and its compaction phase in the stress-strain curve is longer than that in the natural state. Deformation characteristics of granite in the elastic deformation phase are mainly controlled by the elastic modulus. Other than water-chemical coupling effect that may change the elastic modulus of granite, confining pressure may also affect the elastic modulus. Sensitivity of the elastic modulus to chemical solutions has been discussed in analysis on the uniaxial compression test results, so its details are not elaborated. Triaxial compression peak strength results of granite specimens after natural drying and with action of different chemical solutions are shown in Table 5-8.

Triaxial compression test peak strength results of granite specimens Table 5-8

Peak strength(MPa) \ Confining pressure(MPa) \ Hydrochemical environment	5	10	20	30
Natural drying	153.50	190.67	226.23	266.52
pH=2, 0.01 mol/L NaCl Solution	105.28	114.46	155.62	192.23
pH=4, 0.01 mol/L NaCl Solution	110.12	149.55	163.89	210.45
pH=7, 0.01 mol/L NaCl Solution	141.37	164.40	188.52	238.25

(3) Impact of chemical solution on c and φ values of granite

As chemical interaction between the chemical solution and rock will weaken strength parameters of rock, and cohesion and friction angle are both important parameters of strength the rock. Based on results of triaxial compression tests, the impact of water-chemical coupling on cohesion and internal friction angle of granite is studied. Based on relationship between peak intensity and confining pressure, relationship of σ_1-σ_3 is shown in Figure 5-19.

It can be learned from triaxial compression σ_1-σ_3 diagram of granite in Figure 5-19 that peak strength of specimen has a relatively obvious linear relationship with confining pressure and its failure conforms to the Mohr-Coulomb strength criterion. So based on Mohr-Coulomb strength theory, peak strength of granite under different confining pressures is used to calculate its cohesion and friction angle before and after the action of chemical solution. In the Mohr-Coulomb strength criterion, cohesion and internal friction angle of the rock can be calculated below:

$$\varphi = \arcsin \frac{k-1}{k+1} \tag{5-6}$$

$$c = \frac{b(1-\sin\varphi)}{2\cos\varphi} \tag{5-7}$$

where c—cohesion of the rock;
φ—internal friction angle of the rock;
k, b—slope and intercept of the straight line drawn on the σ_1-σ_3 coordinate graph using the least square method.

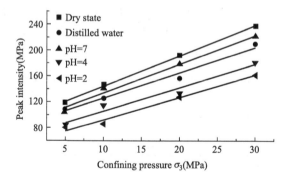

Figure 5-19 Triaxial compression σ_1-σ_3 curve of granite

So based on triaxial compression σ_1-σ_2 relationship diagram of granite in Figure 5-19, equations (5-6) and (5-7) are used to calculate cohesion and friction angle of the granite specimen in the test. The results are seen in Figure 5-20.

It can be seen by analyzing Figure 5-20 that chemical solution has a certain effect on cohesion and internal friction angle of granite. cohesion of granite under natural drying is 33.40 MPa, and after it is soaked in NaCl solution with pH=2 for 60 days, its attenuation is 21.99 MPa or by a range of 34.16%; after action

5 Rock Chemical Damage Induced by Deep Hydrochemical Environment

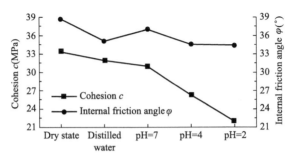

Figure 5-20 Cohesion and internal friction angle of granite after action of different hydrochemical solutions

of other chemical solutions, cohesion of granite specimens also shows attenuation of varying ranges, indicating that the water-chemical coupling effect can significantly alter cohesion of granite.

When compared with cohesion, internal friction angle of granite does not change much after action of chemical solution. Internal friction angle of granite in natural drying is 38.63°, and that of granite is 34.40° and 34.54° after action of NaCl with pH=2 and pH=4. Their drop range is smaller than that of cohesion. It can be thus seen that sensitivity of granite cohesion to water chemical solutions is greater than that of internal friction angle to water chemical solutions. It also shows that cohesion is a structural parameter of the specimen and is largely affected by distribution of joints and cracks and the saturated state. Internal friction angle, on the other hand, is an elastic parameter and is not very sensitive to crack structure and water content of the rock.

5.2.5 Splitting test

(1) Test plan

A standard disc specimen with a size of 50 mm×25 mm is used for the rock specimen used in the split test. Before the test, all specimens are subject to the immersion test treatment other than natural working state. The design conditions and coding of specimens of the split test are shown in Table 5-9. Three specimens are prepared for each operating condition in order to control discreteness of the test results. The test is carried out on WEP-600 screen all-purpose testing machine.

(2) Experimental results and analysis

Based on failure load of each specimen in the test, split strength of the rock is obtained from the following equation:

$$\sigma_t = 2P_{max}/(\pi DH) \qquad (5-8)$$

where σ_t—rock resistance P is failure load of the rock mass, N;

P_{max}—rock mass failure load, N;

D, H—diameter and height of the test piece.

Splitting test conditions and specimen coding   Table 5-9

Coding of work condition	Coding of specimen	Chemical solution	Recycle speed of solution
1	A1-1, A1-2, A1-3	Natural drying	—
2	B2-1, B2-2, B2-3	Distilled water	$v=0$ mm/s
3	B3-1, B3-2, B3-3		$v=200$ mm/s
4	B5-1, B5-2, B5-3		$v=400$ mm/s
5	C5-1, C5-2, C5-3	0.01 mol/L NaCl solution, pH=2	$v=0$ mm/s
6	C6-1. C6-2, C6-3		$v=200$ mm/s
7	C7-1, C7-2, C7-3		$v=400$ mm/s
8	D8-1, D8-2, D8-3	0.01 mol/L NaCl solution, pH=4	$v=0$ mm/s
9	D9-1, D9-2, D9-3		$v=200$ mm/s
10	D10-1, D10-2, D10-3		$v=400$ mm/s
11	E11-1, E11-2, E11-3	0.01 mol/L NaCl solution, pH=7	$v=0$ mm/s
12	E12-1, E12-2, E12-3		$v=200$ mm/s
13	E13-1, E13-2, E13-3		$v=400$ mm/s

Splitting test result   Table 5-10

Specimen coding	Diameter (mm)	Height (mm)	Failure load (KN)	Tensile strength (MPa)	Average tensile strength(MPa)
A1-1	50.38	25.62	21.41	10.57	
A1-2	49.44	25.04	22.63	11.64	12.16
A1-3	49.84	25.73	28.71	14.26	
B2-1	49.51	25.42	25.97	13.14	
B2-2	49.18	24.62	18.96	9.97	11.71
B2-3	49.47	25.10	23.43	12.02	
B3-1	49.65	25.68	18.98	9.48	
B3-2	50.46	24.08	24.93	13.07	11.52
B3-3	49.08	24.02	22.20	12.00	

continued

Specimen coding	Diameter (mm)	Height (mm)	Failure load (KN)	Tensile strength (MPa)	Average tensile strength(MPa)
B5-1	49.54	24.91	19.49	10.06	
B5-2	50.59	25.24	27.22	13.58	11.82
B5-3	49.91	24.63	*10.53	5.46	
C5-1	50.23	25.34	21.54	10.78	
C5-2	50.12	24.49	17.84	9.26	10.04
C5-3	49.37	24.93	19.47	10.08	
C6-1	50.21	25.78	16.80	8.27	
C6-2	49.83	25.55	19.25	9.63	9.25
C6-3	50.76	25.21	19.76	9.84	
C7-1	49.67	24.90	17.09	8.82	
C7-2	49.34	25.71	19.40	9.74	8.94
C7-3	49.97	25.03	16.19	8.25	
D8-1	50.52	25.72	21.13	10.36	
D8-2	50.57	24.08	21.61	10.30	11.06
D8-3	49.66	25.24	22.69	11.53	
D9-1	49.06	25.06	21.13	10.36	
D9-2	50.83	25.49	19.28	9.48	10.49
D9-3	49.70	24.14	20.29	10.77	
D10-1	50.36	25.31	16.35	8.17	
D10-2	49.98	25.26	18.94	9.55	11.90
D10-3	49.57	25.52	25.37	12.78	
E11-1	49.10	25.28	25.58	13.13	
E11-2	50.56	25.39	*11.39	*5.92	10.17
E11-3	49.78	24.69	20.59	10.67	
E12-1	49.97	24.38	21.57	11.28	
E12-2	49.53	25.16	22.96	11.74	10.38
E12-3	49.51	24.79	21.40	11.13	
E13-1	49.12	25.50	25.01	12.72	
E13-2	49.88	25.08	19.08	9.71	10.86
E13-3	50.31	24.77	19.84	10.14	

Note: * stands for abnormal data and is not used for calculation of average tensile strength.

Split test results of each specimen are seen in Table 5-10. It can be seen from Table 5-10 that tensile strength of the granite specimens decreases in varying degrees after the action of hydrochemical solution.

To study the impact of hydrochemical solution on tensile strength of granite, D_t, the damage amount of tensile strength of rocks with action of hydrochemical action is thus defined. The equation is seen below:

$$D_t = 1 - \sigma_{t(t)}/\sigma_{t(0)} \qquad (5-9)$$

where D_t—water-rock interaction damage after water-rock interaction for a time duration of t;

$\sigma_{t(t)}$—uniaxial tensile strength of rock after the action of water and rock for a time duration of t;

$\sigma_{t(0)}$—zero damage initial value of uniaxial tensile strength of rock.

The average tensile strength of the granite specimen at 12.16 MPa in natural drying is taken as zero damage initial value. Damage of the granite tensile strength and its changing trends after the action of different water chemical solutions with different flow rates for 60 days are shown in Table 5-11 and Figure 5-21.

Chemical damage amount of tensile strength of granite specimens under different water-rock interaction environments   Table 5-11

Chemical damage D_t	Natural drying	Distilled water			NaCl solution, pH=2		
		v=0 mm/s	v=200 mm/s	v=400 mm/s	v=0 mm/s	v=200 mm/s	v=400 mm/s
	1	0.963	0.947	0.972	0.826	0.761	0.735
Chemical damage D_t	Natural drying	NaCl solution, pH=4			NaCl solution, pH=7		
		v=0 mm/s	v=200 mm/s	v=400 mm/s	v=0 mm/s	v=200 mm/s	v=400 mm/s
	1	0.910	0.863	0.836	0.979	0.936	0.893

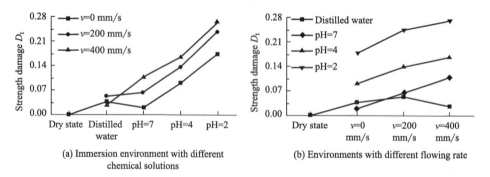

(a) Immersion environment with different chemical solutions

(b) Environments with different flowing rate

Figure 5-21 Tensile strength of granite specimens before and after action of different water-chemical coupling

After analyzing Table 5-11 and Figure 5-21, the following aspects are concluded:

1) It can be learned from Figure 5-21 (a) that average tensile strength of granite specimens immersed in static water decreases with increase of solubility of the chemical solution. Lower pH value of the solution means higher chemical damage value. The damage amount reaches 0.826 after the action of 0.01 mol/L NaCl solution with pH=2. In static water, H^+ ions in the solution penetrate into the granite through gaps and micro-cracks of the rock and react with some of its components so that the originally existing micro-cracks will develop, expand and even go through penetration, and direct cohesion of rock particles can decrease. It is manifested in decline of tensile strength of the rock on the macroscopic level.

2) It can be learned from Figure 5-21 (a) that chemical damage of granite in dynamic water is more significant than that in static water. Under the action of 0.01 mol/L NaCl solution with pH=2 and a circulating flow rate of 400 mm/s, the chemical damage amount is the maximum at 0.735.

3) Figure 5-21 (b) reflects degree of change in tensile strength of granite after action of different flow rates of the same solution. Chemical damage of granite is more sensitive to the flow rate in the NaCl solution and the change range in tensile strength is more significant. In distilled water this trend is not obvious.

4) Damage on tensile strength of granite under the action of chemical solution shows the following trends: for the same water flow environment, higher acidity in the chemical solution means more significant damage; when chemical solution has the same concentration, higher flow rate leads to more obvious damage amount. Under experimental conditions of this study, damage of tensile strength of granite is in the controlling position among all factors due to chemical action and its impact on tensile strength of the rock is greater than that with action of water flow.

5.3 Morphology and porosity development of granite before and after chemical erosion

On the ground of the macro-mechanical test on the chemical failure effect of granite, this paper starts by applying the scanning electron microscopy and electronic energy spectroscopy and other means to gauge and analyze the transformation effect of hydrochemical solution on the granite's macro-and micro-shape, micro-particles and element composition. It tests and measures the granite specimens under different immersion flow rates to obtain the time effect curve of hydrochemical failure time of granite and to analyze quality of the granite specimens, solution pH value, elastic longitudinal wave velocity and solution ion composition under the same immersion time nodes. To conclude, it follows test results of the mineral and compound composition of granite before and after the chemical action and the chemical kinetic theory to probe into chemical failure mechanism during interaction between the rock and the water chemical solution.

5.3.1 Macro-morphology characteristics of granite after chemical erosion

After granite is immersed in different chemical solutions, its surface has been imprinted with corrosion traces of different degrees. Figure 5-22 shows comparison of shape characteristics on the surface of some granite specimens with typical representative significance before and after immersion in chemical solution. Corrosion characteristics of granite specimens are obvious after being immersed in 0.01mol/L NaCl solution (pH=2) for 60 days. Particles on the surface of the specimens that have not been corroded are relatively smooth. After being corroded by chemical solution, particles on the surface become relatively coarse. At the same time, original micro-cracks of the specimens expand and extend; the dense, crack-free granite specimen before immersing also produce obvious cracks on the surface after chemical corrosion.

Although corrosion effect of 0.01 mol/L NaCl solution (pH=4) on granite is not as significant as that of 0.01 mol/L NaCl solution (pH=2), and large-area surface spalling and dissolution is rarely seen, corrosion characteristics on the surface of specimens are also more obvious. Corrosion effect of 0.01 mol/L NaCl solution (pH=7) and distilled water on granite is very limited, and surface shape of the specimens before and after immersion stays basically the same. After analyzing the surface shape characteristics of all granite specimens before and after the action of chemical solution, it can be found that corrosion of the granite specimens under the action of different chemical solutions at different flowing rate is mainly affected by nature of the mineral and the chemical environment (solution concentration, pH value), and the flow rate of the solution is not a controlling factor.

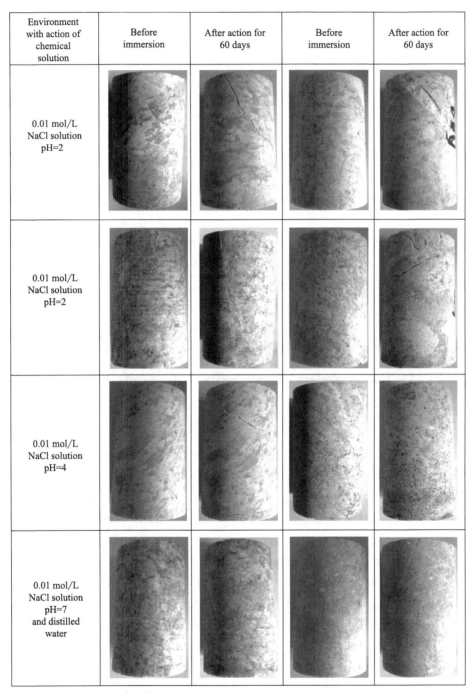

Figure 5-22 The apparent shape characteristics of granite specimens after being immersed in acid chemical solution

Granite's sensitivity towards different solutions varies. Tests can reveal that different chemical solutions have different corrosion effects on granite. In fact, even under the same chemical environment and at the same water flow, corrosion of granite also shows variation and nonuniformity: corrosion effect of some specimens is mainly reflected in propagation and penetration of cracks, and dissolution and spalling of surface particles is not significant; some specimens show no obvious changes in characteristics of cracks, but corrosion phenomenon on the surface is very serious. It can be shown in specimens in the first column of Figure 5-22. It can be thus seen that chemical corrosion has selective differences in the effect of granite. This is because granite is a collection of multiple mineral and different minerals react physically and chemically with the solution, and their reaction mechanisms, methods and results also vary, so the granite shows selectivity of corrosion in the chemical reaction. Moreover, existence of micro-cracks in granites aggravates the chemical corrosion. The granite has significant heterogeneity and anisotropy. Chemical corrosion with selective difference will further reduce continuity and integrity of the granite while the heterogeneity will be strengthened, so its fracture pattern will eventually become more complicated.

5.3.2 Micro-morphology characteristics of granite before and after chemical erosion

The effect of chemical corrosion on the macroscopic shape of surface of granite specimen is more significant. In fact, changes of the macroscopic shape of granites are closely related with change of the microstructure. With the help of SEM scanning electron microscopy analysis methods, morphological characteristics and pore structure distribution of rock particles can be visually observed. In the meantime, XRD element energy spectrum analysis technology can be used to obtain element content of all parts on the surface of rock particles. To study the impact of chemical action on microscopic shape of granites, SEM and elemental energy spectrum test is done on granite specimens before and after action of chemical solutions on QUANTA 200F field emission scanning electron microscope (Figure 5-23).

When carrying out the test on scanning electron microscopy, shape of the microscopic pores and defects on the surface of the granite specimen before and after the chemical action is compared and analyzed, and changes in content of the mineral element content around the pores are quantitatively tested. The observa-

5 Rock Chemical Damage Induced by Deep Hydrochemical Environment

Figure 5-23 QUANTA 200F field emission scanning electron microscope

tion and analysis results of scanning electron microscopy and energy spectrum analysis are shown in Figure 5-24 and Figure 5-25.

SEM spectrum of granites in Figure 5-24 shows that after the granite is subject to different chemical solutions, different degrees of corrosion marks show up, microscopic shape witness specific changes and chemical corrosion on granites by different chemical solutions also shows great variation. The surface of granites before chemical action has a dense structure with extensive cementation surfaces among mineral crystal particles, as is shown in Figure 5-24 (a). Continuity among particles is sound. After action of 0.01 mol/L NaCl solution with pH=2 for 60 days, the surface structure of granites is loose, large crystal grains develop into small clastic particles and many secondary pores are produced (Figure 5-24b); the microscopic appearance of granites after action of 0.01 mol/L NaCl, solution with pH=4, the microscopic shape characteristics are similar to those of the solution with pH=2, as structure is loose and secondary pores are developed, but their corrosion degree is slightly worse; from Figure 5-24 (d) and Figure 5-24 (e), it can be seen that in the environment of 0.01 mol/L NaCl solution with pH=7 and of distilled water, edges and corners of the mineral particles become smooth, and an iota of small particles are produced in some areas. Surface structure of the granite specimen is relatively dense and the secondary pores are not obvious, which indicates that the two chemical solutions have a limited effect on the granite.

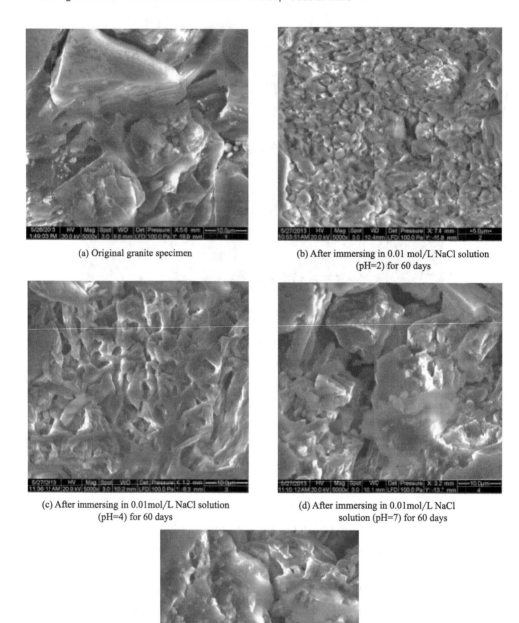

(a) Original granite specimen

(b) After immersing in 0.01 mol/L NaCl solution (pH=2) for 60 days

(c) After immersing in 0.01mol/L NaCl solution (pH=4) for 60 days

(d) After immersing in 0.01mol/L NaCl solution (pH=7) for 60 days

(e) After immersing in distilled water for 60 days

Figure 5-24 Scanning electron microscope images of surface of granite specimens

5 Rock Chemical Damage Induced by Deep Hydrochemical Environment

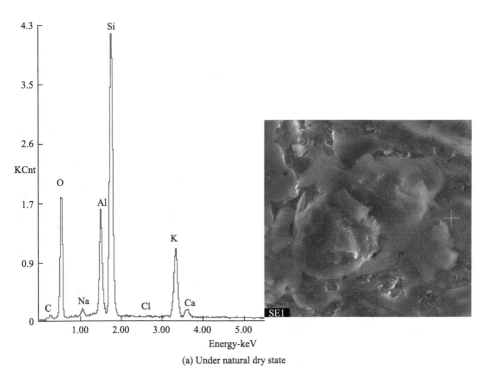

(a) Under natural dry state

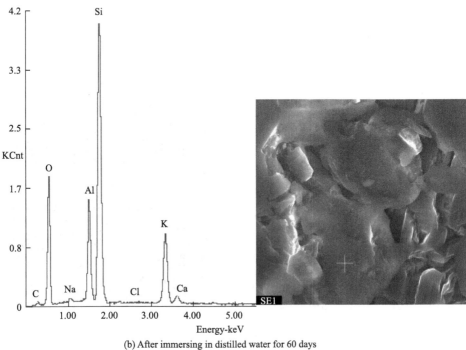

(b) After immersing in distilled water for 60 days

Figure 5-25　Results of electronic energy spectrum analysis of granite specimens（one）

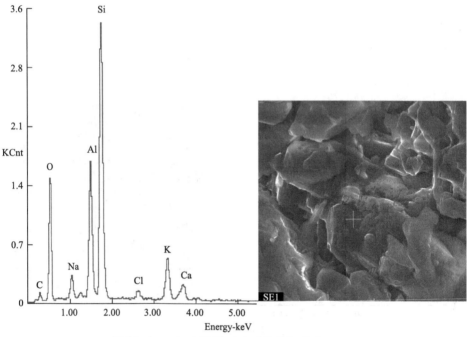

(c) After immersing in NaCl solution(pH=7) for 60 days

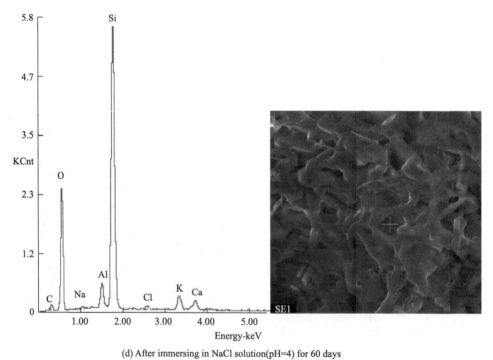

(d) After immersing in NaCl solution(pH=4) for 60 days

Figure 5-25 Results of electronic energy spectrum analysis of granite specimens (two)

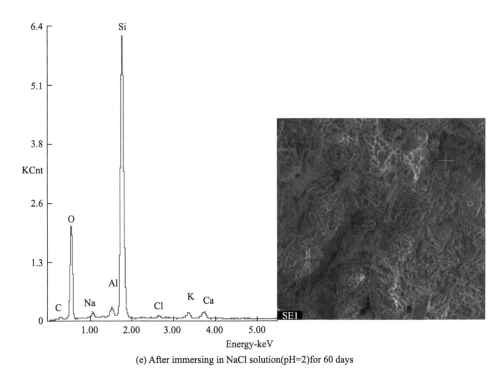

(e) After immersing in NaCl solution(pH=2)for 60 days

Figure 5-25 Results of electronic energy spectrum analysis of granite specimens (three)

As electronic energy spectrum analysis is limited to a small part in a microscopic area, its analysis results may show great randomness. So to avoid errors caused by such random factors, multiple energy spectrum analyses are performed at different locations in different areas of the specimen, average values of masses and atomic numbers of the main elements are obtained and their changes are analyzed. As can be found from Figure 5-25, due to adsorption of granites, content of Cl element in the granite increases to a certain extent after action of NaCl solution; if randomness of the analysis results is not considered, the content of each element does not show much variation after action of 0.01 mol/L NaCl solution with pH=7 and that of distilled water; after action of 0.01 mol/L NaCl solutions of pH=2 and pH=4 for 60 days, the content of Na, Mg, Al, K, Ca and so on in the granite specimen all show decrease to a certain extent.

5.3.3 Impact of chemical solution on porosity of granite

Pore structure of rock alters under impact of external environmental factors. Among external environmental factors, physical and chemical reaction of underground water on rock leads to hydrolysis and dissolution of the rock, which is one of the important factors affecting the pore structure of rock. On the one hand, underground water produces physical effects on rocks such as lubrication, softening, siltification, strengthening of bound water, and scouring and migration. And on the other hand, it has chemical action on rocks covering dissolution, hydrolysis, ion exchange and oxidation-reduction reactions and thus alters mineral structure and composition of rocks on the microscopic basis. As active minerals are removed out of rock, karst caves and corrosion cracks are formed in it, which increase its porosity, affect permeability and pore pressure of the rock and hence alter its macroscopic physical and mechanical properties. The impact of the physical and chemical effects of underground water on the rock pore structure is reflected in the chemical kinetic process of mineral dissolution in the microscopic view, and in changes og porosity of the rock over time in the macroscopic view. Study of change of rock porosity due to the coupling effect of water and rock is necessary to grasp the mechanism of hydrochenical damage of rocks. Porosity of the granite specimens in natural drying or after the action of different chemical solutions for 60 days is measured based on pycnometer method. Three specimens are prepared for each test condition, an porosity is calculated from dry density and particle density of rock based on equation (5-10). Measurement results on porosity of the granite specimens are shown in Figure 5-26.

$$n = \frac{\rho_p - \rho_d}{\rho_p} \times 100\% \qquad (5-10)$$

where n—porosity;

ρ_d—dry density of the rock;

ρ_p—particle density of the rock.

By analyzing results in Figure 5-26, it can be seen that porosity of the granite specimens increases after action of different chemical solutions for 60 days. When compared with natural drying, the dwindling sequence on increase range of granite porosity is distilled water, NaCl solution ($pH = 7$), NaCl solution ($pH=4$) and NaCl solution ($pH=2$).

Based on SEM test results on microshape of granite, the secondary pores of granite specimens increase significantly with expansion of specific surface area and overall loose structure after action of different chemical solutions. One of the macroscopic manifestations of the aforesaid microscopic characteristic changes of granite is increase of porosity. Increase range of porosity is the maximum for

0.01 mol/L NaCl solution (pH=2) after action for 60 days, which is followed by NaCl solution with pH=4, NaCl solution with pH=7 and distilled water the smallest. It indicates that generation of secondary porosity in granite is highly related to concentration of H^+ ions in the aqueous solution. Moreover, Figure 5-26 reveals that after action of the chemical solution, discreteness of porosity measurement results of the granite specimens is more significant than that in natural drying. With heterogeneity of rock aside, it might be related to difference in selection of rocks by chemical corrosion.

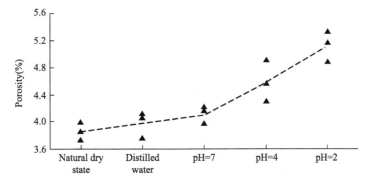

Figure 5-26 Porosity of granite specimens after action in natural drying and different chemical solutions for 60 days

5.4 Time-dependent characteristic test of granite damage process under the action of acidic chemical solution

5.4.1 Testing method

In the granite chemical-seepage coupling macro-mechanics test, the granite specimens are immersed in different chemical solutions for 60 days before the macro-rock mechanics test. To study the mechanism and time-effect characteristics of the chemical-seepage coupling effect on physical indexes of granite, quality of the granite specimens, elastic longitudinal wave velocity and the pH value of the chemical solution are measured at regular intervals during the immersion process. After the immersion test, ion chromatographic analysis is done on samples of the final chemical solution to determine its ion composition and content.

During the test on immersion of chemical solution, it is found that the reaction of granite and chemical solution is obvious in the initial stage of immersion, and then active minerals in the granite produce vast bubbles due to dissolution. Intensity of the chemical reaction gradually decreases with time, and eventually reaction between the rock and chemical solution tends to reach balance. So in the initial stage of the immersion test, a shorter time interval is selected to measure the granite specimens and chemical solutions. When the immersion time is 0.5 d, 1 d, 2 d, 3 d, 5 d, 10 d, 15 d, 25 d, 40 d and 60 d, quality of the granite specimen and pH value of the chemical solutions are measured. When the immersion time is 0.5 d, 1 d, 2 d, 3 d, 5 d, 10 d, 15 d, 30 d, 60 d, elastic wave longitudinal wave velocity is measured on the granite specimens. After the immersion test, specimens of each group of chemical solutions go through quantitative analysis of ion chromatography to determine ion composition and concentration of chemical solution after immersion.

5.4.2 Results and analysis of time-dependent test of rock quality failure

Rock quality is one of the important parameters reflecting basic physical characteristics of rocks. When measuring quality of granite specimens in the immersion test, if they are directly taken from the immersed acid tank and weighed, chemical solution and moisture on the surface of and inside the specimen will have a certain random impact on the results and practical operation is inconvenient; if weighing is carried out after drying, time and process of the normal immersion test will be affected. So after taking the aforesaid factors into consideration, in practical operation of quality measuring, the special inner-outer double-layer design of the acid tank is used and both the inner tank and specimens in it are taken out together to be located in an indoor ventilated environment for 30 minutes. After surface of the specimen is dry, quality measuring is done (Figure 5-27). Other than the specimen that enters quality testing after complete drying on the 60th day, quality of the specimen at other times is measured following the aforesaid method.

Figure 5-28 shows results of quality of granite specimens when observed at different times under action of different chemical solutions. By taking the average quality result of rock specimen and analyzing quality change of the rock, the change rule on time of chemical corrosion and the rock quality is obtained.

It can be learned from Figure 5-28 that the mean quality of each group of specimens decreases to varying degrees after being immersed in chemical solu-

5 Rock Chemical Damage Induced by Deep Hydrochemical Environment

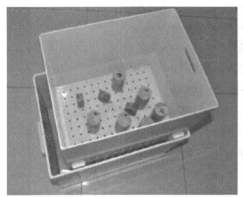

Figure 5-27 Quality measuring of granite specimens under chemical water-rock interaction

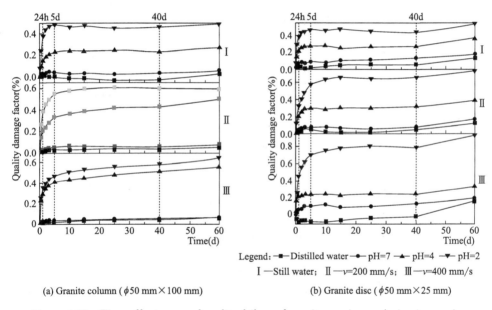

Legend: ■ Distilled water ● pH=7 ▲ pH=4 ▼ pH=2
I —Still water; II —v=200 mm/s; III —v=400 mm/s

(a) Granite column (ϕ50 mm×100 mm)   (b) Granite disc (ϕ50 mm×25 mm)

Figure 5-28 Time-effect curve of quality failure of granite specimens during immersion

tion. To represent degradation degree of granite quality after being immersed in chemical solution, D_m, quality failure factor, is defined below:

$$D_m = \frac{m_o - m_t}{m_o} \quad (5\text{-}11)$$

where D_m—quality failure factor of the rock;
m_o—initial quality of the rock before immersion;
m_t—quality of the rock at a certain time during the immersion process.

Qualiy measured on the 60th day is the result after the specimen goes through complete drying. In equation (5-11), take $t=60$ d and final quality loss rate of the specimen after action of the chemical solution is obtained, which is listed in Table 5-12 and Table 5-13. Quality changes of rock specimens of $\phi50$ mm×100 mm and $\phi50$ mm×25 mm basically follow identical rules.

Changes of rock quality under action of chemical solutions show significant time effect. When the immersion time is less than 5 days, quality of the granite specimens drastically reduces, which is notably so for reduction range of quality within 24h as loss of rock quality during this time accounts for about 50%-70% of the final loss. After immersion for 1-5 days, reduction range of quality gradually dwindles. When the immersion time is longer than 5 d, quality value of the rock basically tends to be stable. Moreover, after drying and removal of moisture, quality of the specimen measured after immersion for 60 days has a certain degree of decrease when compared with the previous measurement.

Quality loss rate of $\phi50$ mm×100 mm granite specimen after action of chemical solution ($t=60$ d) Table 5-12

	0.01 mol/L NaCl, pH=2			0.01 mol/L NaCl, pH=4		
$D_t(\%)$	$v=0$ mm/s	$v=200$ mm/s	$v=400$ mm/s	$v=0$ mm/s	$v=200$ mm/s	$v=400$ mm/s
	0.49	0.59	0.64	0.27	0.50	0.56
	0.01 mol/L NaCl, pH=7			Distilled water		
$D_t(\%)$	$v=0$ mm/s	$v=200$ mm/s	$v=400$ mm/s	$v=0$ mm/s	$v=200$ mm/s	$v=400$ mm/s
	0.06	0.08	0.06	0.04	0.05	0.07

Quality loss rate of $\phi50$ mm×25 mm granite specimen after action of chemical solution ($t=60$ d) Table 5-13

	0.01 mol/L NaCl, pH=2			0.01 mol/L NaCl, pH=4		
$D_t(\%)$	$v=0$ mm/s	$v=200$ mm/s	$v=400$ mm/s	$v=0$ mm/s	$v=200$ mm/s	$v=400$ mm/s
	0.53	0.74	0.93	0.35	0.41	0.33
	0.01 mol/L NaCl, pH=7			Distilled water		
$D_t(\%)$	$v=0$ mm/s	$v=200$ mm/s	$v=400$ mm/s	$v=0$ mm/s	$v=200$ mm/s	$v=400$ mm/s
	0.15	0.17	0.19	0.14	0.12	0.17

Other than acquiring significant time-effect characteristics, changes of rock quality in the experiment are closely related to such factors as chemical solution composition, concentration and flow rate of immersion solution. From Table 5-12 and Table 5-13, it can be seen that quality degradation characteristics of granite follow the following rules: under the same flow rate, lower pH value of the chemical solution, means higher quality loss rate. Under the action of 0.01 mol/L NaCl solution with pH=2 and pH=4, quality of granite is significantly reduced, but it is not obvious at pH=7 and in distilled water; flow rate of the solution also has an important impact on change of quality. When immersed in the same chemical solution, higher flow rate of solution means higher mass loss rate. During the test observation time spanning 60 days, reduction of quality of the specimen is the maximum for 0.01 mol/L NaCl solution of pH = 2 (v = 400 mm/s), which is about 0.93%; that in distilled water (v=0 mm/s) was the minimum at about 0.05%; By comparing quality changes at v=0 mm/s and v=200 mm/s, v=200 mm/s and v=400 mm/s, it can be seen that under action of the same chemical solution, reduction in rock quality will increase significantly when static water immersion is converted to dynamic water; when low flow rate is switched to a high flow environment, quality reduction is not as obvious as that earlier. Overall put, although quality changes of specimens sized ϕ50 mm×100 mm and ϕ50 mm×25 mm are different in values, their changing rules are basically identical.

5.4.3 Time-dependent velocity of elastic longitudinal wave

Propagation speed of ultrasonic waves in rock masses is closely related to physical and mechanical properties of rock masses. Dynamic elastic modulus of rocks, Poisson's ratio and other parameters can be determined by testing propagation condition of longitudinal and transverse wave speeds in the rock; development degree of pores and fractures in the rock, water content and stress state will all affect propagation speed of ultrasonic waves in the rock. So wave speed is an important piece of information reflecting comprehensive physical properties of rocks. During immersion test of granite in chemical solutions, the acoustic wave tester is used to test acoustic wave of granite specimens at different times to study changing rules of time effect of velocity of elastic longitudinal waves of granite under the action of chemical corrosion. Based on results of the acoustic wave test, the curve on velocity of the average elastic longitudinal wave of the granite specimens under the action of chemical corrosion with time is delineated.

From Figure 5-29, it can be found that v_p-t curve of the granite specimens during immersion process in NaCl solution with pH=2 and pH=4 shows the fol-

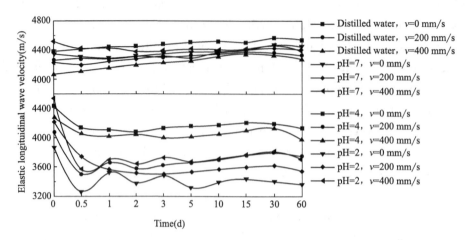

Figure 5-29 v_p-t curve of granite specimens under chemical corrosion

lowing change rules:

v_p, the longitudinal wave velocity, first shows decrease to varying degrees and then longitudinal wave velocity of the specimen increases or decreases with time. Change tendency of wave velocity of some specimens shows a sinusoidal wave pattern. For example, in 0.01 mol/L NaCl solution with pH=2 ($v=0$ mm/s, $v=400$ mm/s), amplitude of wave change gradually decreases with time. When $t>10$ d, the v_p-t curve gradually stabilizes. The v_p-t curve of granite under the action of NaCl solution with pH=7 and distilled water shows fluctuation to a certain extent at the initial stage of immersion, but it shows an overall growing tendency. When $t=90$d, v_p of all specimens after drying decreases.

Under water-rock chemical coupling, the time-effect change mechanism of v_p, the longitudinal wave velocity of granite specimens, is more complicated. By combining with erstwhile research conclusions and results of this test, the following analysis is made:

(1) Hions in NaCl solution with pH=2 and pH=4 react with active oxides in the granite so that it can corrode rapidly and its porosity increases. At the same time, softening or swelling of soluble minerals leads to reduction of wave velocity. From test results on quality of the specimen and pH value of the solution, it can be seen that corrosion effect of the chemical reaction decreases rapidly with time. In the meantime, water absorption increases saturation of the rock and uniformity of the rock. The aforesaid factors lead to increase of elastic longitudinal wave velocity. Microscopic distribution effect on change of rock elastic wave velocity trig-

gered by selective difference of chemical corrosion effect and variation of saturability leads to changing trend of wave velocity, which presents a sine-like fluctuation. The v_p-t curve under the action of circulating water chemical solution in Figure 5-29 shows more significant fluctuation, for when compared with the static water environment, dynamic water environment of the circulating water increases water absorption role of the rock as well as randomness and uncertainty of water-rock chemical reaction and hence change trend of wave velocity is more obvious. As pH value of the immersion solution approaches stability, the granite specimens gradually reach chemical saturation and the elastic wave velocity of rocks also tends to be stable.

(2) Under the environment of NaCl solution with pH=7 and distilled water, chemical reaction effect is very limited. Pores are gradually saturated after aqueous solution enters the rock, which is manifested in gradual increase of the longitudinal wave velocity. When $t=90$ d, all specimens are dried with deprived moisture and reduction of saturation. Measured wave speeds all show decrease to different degrees. Through aforesaid analysis, it can be concluded that although saturation of rock increases after water-rock chemical action, the chemical corrosion action causes activation, migration and dissolution of rock particles through reaction with chemical components in water and active minerals of the rock, resulting in increase of porosity and heterogeneity of the rock and worse physical and mechanical properties of the rock. It eventually reduces the final v_p value. When chemical corrosion effect is not significant, major factors affecting elastic wave velocity cover saturation of rock mass and microscopic distribution of fluid within scale of the specimen.

D_P, damage factor of longitudinal wave velocity of the rock after chemical corrosion is defined as follows:

$$D_P = \frac{v_P - v_P'}{v_P} \tag{5-12}$$

where D_P—damage factor of velocity of the longitudinal wave of rock;

v_P—velocity of the initial longitudinal wave before the rock is immersed;

v_P'— final velocity of longitudinal wave of the rock after water-rock chemical reaction.

Damage factors of velocity of longitudinal waves of all granite specimens are shown in Table 5-14 after calculation.

Damage factors of velocity of elastic longitudinal wave of granite specimens
after action of chemical solutions   Table 5-14

Chemical solution		v_P(m/s)	$v_P{'}$(m/s)	D_P(%)
0.01 mol/L NaCl pH=2	v=0 mm/s	3728	3218	13.68
0.01 mol/L NaCl pH=2	v=200 mm/s	4064	3405	16.22
0.01 mol/L NaCl pH=2	v=400 mm/s	4384	4569	18.60
0.01 mol/L NaCl pH=4	v=0 mm/s	4280	3987	6.84
0.01 mol/L NaCl pH=4	v=200 mm/s	3933	3598	8.51
0.01 mol/L NaCl pH=4	v=400 mm/s	4216	3833	7.10
0.01 mol/L NaCl pH=7	v=0 mm/s	4194	4296	−2.44
0.01 mol/L NaCl pH=7	v=200 mm/s	4079	4254	−4.30
0.01 mol/L NaCl pH=7	v=400 mm/s	4367	4243	2.84
Distilled water	v=0 mm/s	4235	4388	−3.61
Distilled water	v=200 mm/s	4109	4184	−1.83
Distilled water	v=400 mm/s	3926	4136	−5.36

5.4.4 Time-dependent pH values of chemical solutions

In corrosion effect of chemical solution on rock, mineral composition and content in the rock are the internal factors that play a decisive role while concentration, composition and pH value of the chemical solution are important external factors. Change of pH values of chemical solution with time can intuitively reflect reaction between mineral composition in the rock specimen and chemical solutions. Different reaction intensity and time will have different degrees of impacts on physical and mechanical properties of the rock. So in the immersion test of granite specimens in chemical solutions, pH value of the chemical solution is measured at different time and time-effect characteristics of pH value of chemical solutions in water-rock chemical reaction is studied. Based on test results, curve on changes of pH values of different chemical solutions with time is seen in Figure 5-30.

It can be learned from Figure 5-30 that as chemical reaction with the granite is not obvious, pH value of the 0.01 mol/L NaCl solution with pH=7 has been relatively stable at different flow rates during the 60-day-long test time, and no significant fluctuation is shown. pH value fluctuates between 6.823 to 7.174 near the neutral value of 7. The pH values of 0.01 mol/L NaCl solutions with

5 Rock Chemical Damage Induced by Deep Hydrochemical Environment

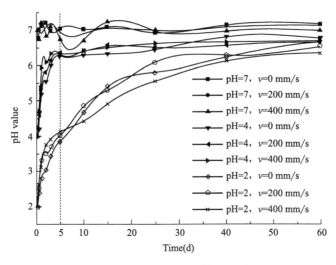

Figure 5-30 Curve on changes of pH value of different chemical solutions with time

pH=2 and pH=4 show an upward trend with time, and eventually tend to be weakly acidic and neutral. When compared with changes of pH value of the 0.01 mol/L NaCl solutions with pH=2 and pH=4 at the same flow rate, it is found that when immersion time $t<5$ d, pH value of the NaCl solution with pH=4 shows a more significant rising trend, and when $t>5$ d, the pH value has been basically stable. This may be ascribed to that molecular weight of hydrochloric acid in NaCl solution with pH=4 is much less than that of NaCl solution with pH=2. When quality of granite is similar, hydrochloric acid molecules in NaCl solution with pH=4 are consumed in no time, as is manifested in rapid surge of pH value. After almost all the hydrochloric acid is engaged in the reaction, pH value of the solution is stable. By analyzing pH curve of NaCl solution with the same concentration but different flow rates, it can be seen that under the circulating water environment, minerals corroded on the surface of granite are taken away by the aqueous solution while new active minerals constantly react with hydrochloric acid so that chemical reaction between the granite and hydrochloric acid is more completed and violent, as is manifested in faster water flow rate and the more significant rise in pH value of the solution.

In summation, due to consumption in reaction of active minerals of granite and certain volatility of hydrochloric acid, pH values tend to be neutral after a time span of 60 days ($t=60$ d) for NaCl solutions with pH=2 and pH=4. The interaction between water and rock is mainly physical and hydraulic and chemical corrosion is very weak.

5.4.5 Variation of ion concentration in chemical solution

The chemical corrosion effect is an important factor in interaction between hydrochemical solution and the rock. After hydrochemical solution interacts with the rock, change of macroscopic physical and mechanical properties of the rock is related to microscopic effect of chemical corrosion of the hydrochemical solution. Active minerals in the rock will undergo a series of chemical reactions under the action of different chemical solutions, which causes certain elements in the minerals to be precipitated in an ionic state, and increases salinity of the aqueous solution. So quantitative analysis on composition of the hydrochemical solution after interacting with the granite is used to deduce type and extent of chemical reaction between the rock and the chemical solution. To explore the reaction mechanism between different hydrochemical solutions and granite and clarify minerals and compound components in the granite that are involved in the hydrochemical reaction, all chemical solutions in the static water immersion environment are specimend and tested by ion chromatography after the immersion test for 60 days. After testing, metal cations in the specimen mainly include Na^+, Ca^{2+}, Mg^{2+}, Al^{3+}, Fe^{3+}, K^+ and so on. Detailed results of the test are seen in Figure 5-31. As the immersion solution contains a certain amount of NaCl, Na^+ precipitated from the granite due to chemical reaction is much less than that in the original immersion solution, so changes on content of Na^+ are not discussed.

Figure 5-31 reveals that concentrations of newly-precipitated ions in chemical solutions are significantly different when immersed in different chemical solutions. After the granite specimen is immersed in 0.01 mol/L NaCl solution with pH=2, the feldspar and calcite in the granite reacted with H^+ in the solution in an acidic environment and hence Ca^{2+}, Mg^{2+}, Al^{3+}, Na^+, K^+ and so on are generated. Due to variation in sensitivity of various mineral components and minerals to chemical environment, concentration of each ion in the same chemical solution drastically varies. For example, after immersion in 0.01 mol/L NaCl solution with pH=2 for 60 days, concentration of Ca^{2+} reaches 558.2 ppm and that of Al^{3+} and Mg^{2+} are 24.3 ppm and 7.4 ppm respectively. After immersion in 0.01 mol/L NaCl solution with pH=4 for 60 days, a certain amount of corresponding ions are also produced in the immersion solution. Although concentration of each ion is less than that in 0.01 mol/L NaCl with pH=4, concentration

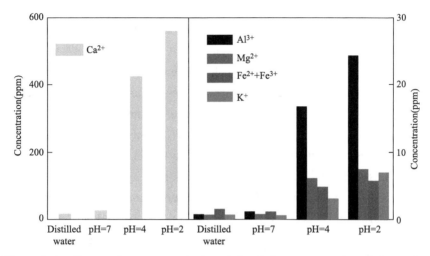

Figure 5-31 Test results on concentration of chemical components in each immersion solution after immersion of granite for 60 days

value of ion is still quite considerable with that of Ca^{2+} and Al^{3+} reaching 423.6 ppm and 16.7 ppm respectively, indicating that reaction between active minerals and H^+ is also very active in the hydrochemical environment with pH=4. After granite is immersed in 0.01 mol/L NaCl solution with pH=7 and in distilled water, concentration of Ca^{2+}, Mg^{2+}, Al^{3+}, K^+ plasma increases slightly. It is because that various minerals in granite are difficult to engage in chemical reaction in a neutral hydrochemical environment and only an an iota of hydrolysis reaction of carbonate minerals and silicate minerals shows up. Results show that among factors that affect concentration of ion precipitation in the solution, pH value of the chemical solution is still the controlling factor.

5.5 Mineral composition change and chemical damage mechanism during rock-groundwater interaction

5.5.1 Variation of mineral composition of granite

Mineral composition of granite is one of the major factors affecting mechanical properties of granite and different mineral components have different sensitivity to the hydrochemical environment. To explore impact of hydrochemical impact on the microscopic mineral components of granite and the resulting changes in physical and mechanical properties, mineral composition and changes of granite specimens before and after immersion in different chemical solutions are measured through X-ray diffraction phase qualitative and semi-quantitative tests. X-ray diffraction analysis is an indispensable conventional method of material structure analysis in mineral crystal structure analysis, mineral identification and research. When X-rays enter crystals, they would engage in various interactions with atoms in the crystal. The coherent scattered X-rays will cause superimposed secondary X-rays that will form diffraction due to periodic impact of atomic arrangement in the crystal. As different crystals have variations in atomic types, location and lattice constants, diffraction directions and intensity will thus vary. For this research and calculation, the atomic positions and lattice constants of matter can be discussed. Correlated tests are performed in the Microstructure Analysis and Testing Center of Peking University (test report number: 130606S01), and analysis results are seen in Table 5-15.

Contents of main mineral components of granite specimens before and after action of chemical after 60 days of immersion Table 5-15

Specimen coding	Percentage of content of minerals(%)				
	Quartz	Plagioclase	Mica	Calcite	Microline
No. 1	42	49	6	3	—
No. 2 (pH=2, 0.01 mol/L NaCl solution)	59	20	12	—	9
No. 3 (pH=4, 0.01 mol/L NaCl solution)	52	33	9	—	6
No. 4 (pH=7, 0.01 mol/L NaCl solution)	46	47	5	2	—
No. 5 (Distilled water)	40	52	7	1	—

Test results reveal that main mineral components of granite in Sanshandao Gold Mine at-600 m under natural conditions cover quartz, plagioclase, mica, and calcite. Based on relevant mineralogical and hydrogeological data, dissolution and corrosion reactions of common minerals in granite in an acidic hydrochemical environment is seen below:

(1) Components of plagioclase form a continuous series among albite, potash feldspar and anorthite. Under acidic conditions, the dissolution equation can be written as:

Potash feldspar: $KAlSi_3O_3 + 4H^+ \longrightarrow Al^{3+} + 3SiO_2 + 2H_2O + K^+$

Albitite: $NaAlSi_3O_3 + 4H^+ \longrightarrow Al^{3+} + 3SiO_2 + 2H_2O + Na^+$

Anorthite: $CaAlSi_3O_3 + 8H^+ \longrightarrow 2Al^{3+} + 2SiO_2 + 4H_2O + Ca^{2+}$

(2) In an acidic environment, calcite in granite is prone to react with H^+ in the solution:

$$CaCO_3 + 2H^+ \longrightarrow Ca^{2+} + H_2O + CO_2$$

(3) Quartz can undergo weak hydrolysis reaction when it is in contact with water:

$$SiO_2 + H_2O \longrightarrow H_4SiO_4$$

(4) An iota of mica in granite also reacts with acid:

$$KAl_3Si_3O_{10}(OH)_2 + 10H^+ \longrightarrow 3Al^{3+} + 3SiO_2 + 6H_2O + K^+$$

In the aforesaid reactions, reactions of plagioclase and calcite in acidic environment are very significant and the dissolution and chemical reaction of quartz is not obvious. Based on X-ray diffraction test results, under action of the NaCl solutions with pH=2 and pH=4, content of plagioclase in the granite is significantly reduced and the calcite is almost completely dissolved. Microcline ($KAlSi_3O_8$), a new mineral, is formed during the reaction. Due to weak reaction between quartz and mica, material consumption is teeny and content in granite has increased to a certain extent. The acidic hydrochemical environment has a significant modification effect on microscopic mineral composition of granite. In other chemical environments (NaCl solution with pH=7, and distilled water), mineral content stays basically unchanged after eliminating heterogeneous factors of the materials.

5.5.2 Chemical composition variation of granite specimen before and after action of chemical solution

X-ray diffraction analysis technique can be used to obtain quantitative results on changes in mineral composition of granite before and after action of the chemical solution, but chemical formula of minerals in the granite is not fixed and the same mineral may be composed of many different end members, as end-member components of plagioclase include albite, potash feldspar and anorthite and content of the three end-member components cannot be completely determined. If X-ray diffraction test results are directly used in subsequent quantitative calculations and analysis, the results will have a certain degree of uncertainty.

Based on the aforesaid issues, X-ray fluorescence spectroscopy other than X-ray diffraction mineral analysis is also conducted on all granite specimens in order to test elements and chemical phases in the specimens. Based on *Test Report of Microstructure Testing and Analysis Center in Peking University* (Report No.

130606S02), quantitative analysis results of X-ray fluorescence spectroscopy is shown in Table 5-16.

Contents of main chemical elements and compounds in granite specimens before and after action of chemical solution after immersion for 60 days   Table 5-16

Specimen coding	Percentage on quality and content of elements and chemical compounds(%)				
	No. 1 Natural drying	No. 2 pH=2 NaCl solution	No. 3 pH=7 NaCl solution	No. 4 pH=7 NaCl solution	No. 5 NaCl solution
C	9.38	12.2	10.89	7.65	9.58
N	0.314	0.556	0.517	0.432	0.253
Na_2O	2.94	0.618	2.40	3.32	4.02
MgO	0.180	0.105	0.272	0.327	0.266
Al_2O_3	16.4	9.03	12.4	19.2	14.8
SiO_2	61.4	72.1	66.8	59.5	64.1
Cl	0.743	0.995	0.469	0.339	0.0926
K_2O	3.47	2.73	3.80	4.86	3.66
CaO	3.52	0.473	0.833	2.76	1.78
TiO_2	0.261	0.24	0.187	0.343	0.109
Fe_2O_3	0.877	0.553	0.754	0.693	1.04
Other	0.515	0.441	0.660	0.576	0.2994

The aforesaid research results reveal that when granite is exposed in an acidic hydrochemical environment, the plasma concentration of Ca^{2+}, Mg^{2+}, Na^+, K^+ will increase over time in the aqueous solution. Test results of fluorescence spectra of chemical elements listed in Table 5-16 can verify this phenomenon. When combining with related chemical kinetics theories, a series of chemical reactions occur between metal oxides in granite and hydrochloric acid, which ultimately leads to decrease of Al_2O_3, Na_2O, MgO and other compounds in the granite and rise of concentration of related ions in the chemical solution under an acidic chemical environments. By analyzing data in Table 5-16 and referring to test results of concentration of solution ion, it can be found that different compounds in granite have different sensitivity to acid. Reaction between Al_2O_3 and CaO and hydrochloric acid is very active and concentration of related ions in chemical solution after reaction is also high. Content of Na_2O and other active compounds that react with acid also decreases to varying degrees. In the hydrochemical environment of NaCl solution with pH=7 and in distilled water, significant chemical reaction is not prone to occur and content of compounds in granite basically stays the same.

5.5.3　Rock damage mechanism induced by the water-chemical interaction

Due to chemical imbalance between rocks and minerals, underground water inevitably undergoes various physical and chemical interactions with rock media during the migration process. They cover dissolution, precipitation, hydrolysis, adsorption and oxidation-reduction. The macromechanical effect of rocks under water-chemical coupling effect is a process during which change of microstructure leads to that of its macromechanical properties depending on composition and chemical properties of aqueous solution, temperature and flow rate, mineral composition of rocks, development of joints and fractures, hydrophilicity and permeability. Impacts of several hydrochemical impacts on rocks are analyzed below:

1) Dissolution. Underground water and minerals in the water-bearing rock mass media constitute a geochemical system containing solid and liquid phases. Due to imbalance between solid and liquid phases in the system, some minerals in the aquifer will dissolve into the aqueous solution. Degree of difficulties in immersion in water by commonly seen minerals follows the following sequence as rock salt, gypsum, calcite, dolomite, olivine, hornblende, plagioclase, potash feldspar, biotite and quartz. Granite used in the test contains plagioclase, mica, quartz, calcite and so on and various mineral components engage in different degrees of dissolution with water. Due to dissolution of the rock, some mineral particles are deprived and the structure turns loose and fragile, resulting in changes of physical and mechanical properties.

2) Precipitation. During dissolution of rocks, dissolution of one mineral may lead to precipitation of another mineral due to the different solubility of minerals. Moreover, chemical interaction between rock and water may generate insoluble salt, and when sediment adheres to the surface of rock particles or defects such as cracks, pores, and fractures, a certain impact will be imposed on mechanical properties of the rock. As sediments cover around pores of the rock, surface structure of the rock becomes dense and overall porosity of the rock mass is reduced, which may produce positive mechanical effects and prove beneficial to mechanical properties of the rock.

3) Adsorption. Adsorption refers to a process during which solutes in water are adsorbed into the solid surface through surface action. Solid matters that dissolve ions in the adsorption solution are usually called adsorbents. Dissolved components in underground water that is moving in the rock are always in contact with surface of adsorbents and constantly engage in absorption so that chemical composition of the rock and underground water witnesses changes accordingly.

Adsorption can be divided into the following two types namely physical adsorption that mainly depends on intermolecular forces with fast adsorption speed and may cause structural changes such as cracks and pores on the rock surface, thereby affecting physical and mechanical properties of the rock; chemical adsorption, ionic composition in adsorbents in the crystalline framework will witness change which will affect size and gap of the nodes in the crystalline framework to a certain extent, but overall crystalline framework will stay the same. Gravitational strength between ions in chemical adsorption may be subject to changes or even crystal change, which will affect physical and mechanical properties of the rock.

4) Redox effect. Redox reaction refers to chemical reaction in which electrons transfer emerges between reactants and products. The process during which a substance is deprived of electrons is defined as oxidation, and the process during which a substance gains electrons is defined as reduction. For mechanical properties of rocks, oxidation-reduction has both positive and negative effects. When the rock is in neutral water or acidic water, the dissolved oxygen in the environment will oxidize with cations in rock minerals and cements before forming an oxidation layer on the rock surface that can inhibit further erosion on the rock by chemical solution. So it is generally held that the impact of oxidation on rock mechanics is positive while that of reduction is negative.

Taking Fe^{2+} that widely distributed in minerals or fillings as an example, it illustrates the dual impact of redox reaction on physical and mechanical properties of rocks. The low-valent iron ion Fe^{2+} can be subject to both oxidation and reduction reactions depending on composition and nature of the surrounding chemical solution Fe^{2+} is located. When Fe^{2+} is engaged in oxidation reaction in a neutral water environment to form $Fe(OH)_3$ precipitation, this process on rock properties is a positive impact. If Fe^{2+} is eroded by acid solution, the iron-containing minerals are oxidized in water and the volume expansion leads to failure rock structure. The mechanical effect of this process is negative. In weak acid and aqueous environment with high salinity, the dissolution or hydrolysis between water and rock is dominant while the adsorption effect is weak; in neutral and aqueous environment with low mineralization, Fe^{2+} ions in rock minerals will adsorb SiO_2 in the neutral aqueous solution and will precipitate on the surface of the rock, notably in the tip of the fracture. Adsorption between water and rocks is dominant and dissolution is weakened. So the dual impact of redox on properties of rocks mainly depends on chemical properties and composition of the aqueous solution.

5) Carbonation. When the aqueous solution is acidic or alkaline, failure on various rocks will speed up. For example, carbonate in water rich in CO_2 will be converted into bicarbonate due to carbonation. Solubility of bicarbonate is higher than Carbonate by dozens of folds. Its equation of conversion is seen below:

$$CaCO_3 + H_2O + CO_2 \longrightarrow Ca[HCO_3]_2$$

Carbonic acid is formed with combination of CO_2 in the aqueous solution with water. CO_3^{2-}, its carbonate, is prone to react with Na^+, K^+, Ca^{2+}, Mg^{2+}, Al^{3+} and other cations in rock minerals that then form soluble carbonates, increase degree of mineral dissociation and hence accelerate other chemistry. Taking potash feldspar as an example, equation for interaction with carbonic acid is as follows:

$$4KAlSi_3O_3 + 4H_2O + 2CO_2 \longrightarrow Al_4(Si_4O_{10})(OH)_3 + 8SiO_2 + 2K_2CO_3$$

With chemical combination of ions in potassium feldspar with CO_3^{2-}, K_2CO_3 takes form and dissolves in water. Precipitated SiO_2 colloid is partly dissolved in water or gelled into opal, and, kaolinite, a new mineral, is formed at the same time. It can be seen that carbonation also affects physical and mechanical properties of rocks to a certain extent.

By and large, due to chemical imbalance between underground water and rock media, the irreversible thermodynamic process between underground water and rock is produced. Various chemical interactions between rock and underground water lead to damage and failure of microstructure of rock mineral components and generation of new minerals and components; with deterioration of structure of rock particles and of cement structure and increase of pores, rock becomes loose and fragile and its mechanical properties alter, its strength drops and its deformation increases. Coupling between the mineral composition and structural characteristics of the rock (pore cracks and so on) and composition and properties of the hydrochemical solution around the rock mass jointly determine failure mechanism of the rock in water-rock interaction. Microscopic composition and microstructure of the rock is affected by hydrochemical failure that will alter physical and mechanical properties and stress state of the rock mass.

6

Mining Optimization and Stability Analysis of Stopes

6.1 Optimization of deep mining method

6.1.1 Deep mining method

The main deep mining methods of Sanshandao Gold Mine include point-pillar upward layering and backfilling mining method, room-pillar alternating upward layering and backfilling mining method and drift upward layering and backfilling mining method.

Point-pillar upward layering and backfilling mining method is the main mining method of Sanshandao Gold Mine, which has some problems in deep mining, such as poor stope safety, large loss and low production capacity. Point-pillar mechanized upward layering and backfilling method increases the possibility of roof caving because of its long stope length, large stope span and large empty roof area. In the mining process, a mining combination is responsible for mining more ore, and the exposure time of stope roof is long, which increases the risk. Meanwhile, due to the need to leave a large number of point-pillars to support the stope roof, resulting in a large loss rate, and the hanging wall roof protection mine reserved to ensure safety also caused a considerable loss rate. The ore loss and resource loss caused by the combination of the two are quite serious.

The mining exposed area and roof sedimentation of drift upward layering and backfilling mining method are relatively small, which is beneficial to the ground pressure control of stope and has high safety. Classifying tailings are used for backfilling. When conditions are met, the excavated waste rock can be transported into the stope for backfilling. In order to ensure the safety of two-step drift stoping operation, the quality of backfilling and top connection shall be strictly controlled in one-step mining and backfilling operation.

553 stope is arranged according to the panel, and adopts the room-pillar alternating upward layering and backfilling mining method. The panel adopts the mining scheme of separating one mining one, and the mining is carried out in two steps, with one-step mining followed by two-step mining, and one-step mining is ahead of two-step mining by section. Different from the general scheme of separating one mining one, the one-step mining and two-step mining in each stope are changed. When the same stope is mined in different sections, there is one-step mining and two-step mining, which are alternating, while adjacent sections are divided into one-step mining and two-step mining. The whole panel mining is one-step mining and two-step mining alternately. The method adopts panel mining, which greatly improves the production capacity, has high mining efficiency, and has lower blasting cost and good safety, but the overall mining and shearing cost is higher.

6.1.2　Deep mining method optimization

In order to comprehensively, fully and fairly compare the applicability of the three schemes used for deep mining in Sanshandao in the process of deep ore body mining, and the advantages of the three mining methods in improving the production capacity, production efficiency and safety of mines, respectively, the ore body in 553 stope was simulated by using the drift upward layering and backfilling method, the room-pillar alternating panel upward layering and backfilling method and the point-pillar upward layering and backfilling method, and the technical and economic indexes of each method were calculated, and their advantages and disadvantages were analyzed. The schematic diagrams of the three mining methods are shown in Figure 6-1 to Figure 6-3. The occurrence state and characteristics of the ore body in the 553 panel are representative in the deep of Sanshandao Gold Mine. The strike length of the ore body is 150 m, the height of middle part is 45 m, and the average thickness is 29 m. The middle part of the ore body is thick and the two wings are thin.

For the drift upward layering and backfilling mining method, it is divided into 4 sections, each section has 3-4 layers, and the height of each layer is 2.5-3 m. The roof control height is 4 m, the backfilling height is 2.5 m, the stope is arranged along the strike, and the approach size is 5 m×4 m. No bottom pillar

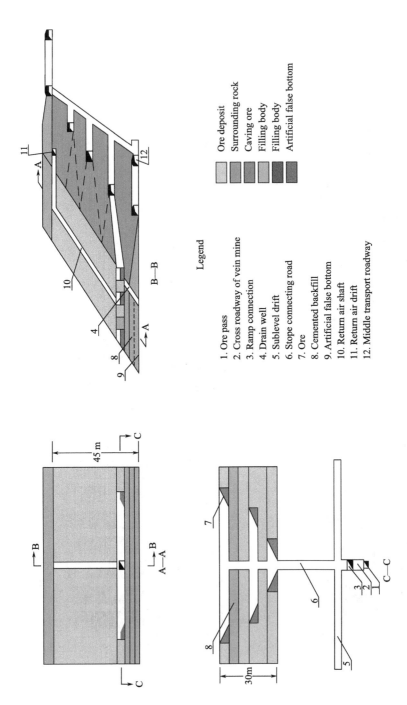

Figure 6-1 Schematic diagram of drift upward layering and backfilling mining method

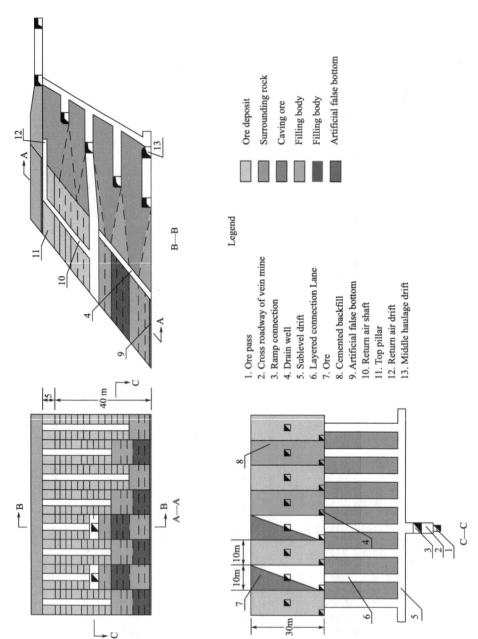

Figure 6-2　Schematic diagram of room-pillar alternating upward layering and backfilling mining method

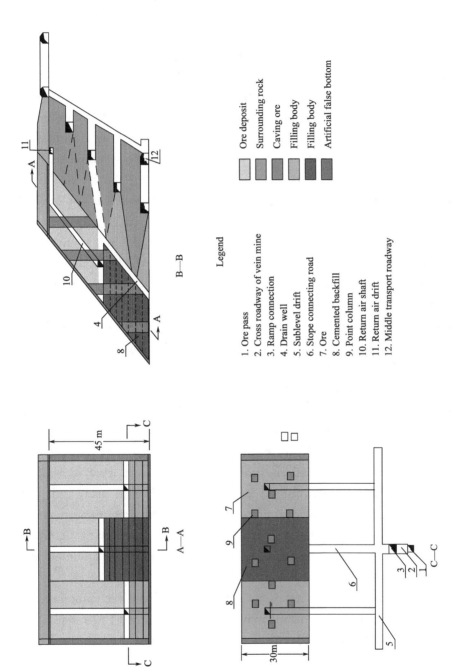

Figure 6-3 Schematic diagram of point-pillar upward layering and backfilling mining method

is left at the floor, the first layered mining height is 4 m, and then a reinforced concrete false floor is built on the floor. Each section is connected with the ore body through the connecting roadway of the layered stope, and the upper layered mining combination is formed by the lower layered mining combination after capping. There are two types of stope layout for the room-pillar alternating upward layering and backfilling mining method: in the the middle ore body with a large thickness, the stope is vertical to the ore body, the stope length is the horizontal thickness of the ore body, the mining area of a single stope is not more than 600 m^2, and the height is 45 m in the middle section; in the thin part of two wing ore body, the stope is arranged along the strike, and the length is the length of the ore body and the width is the width of the ore body. The mining area of a single stope is not more than 600 m^2, and the height is 45 m in the middle section. According to the above requirements, the panel is divided into 8 stopes No. 1-No. 8, among which No. 2-No. 7 stopes are arranged vertically along the ore body strike, while No. 1 and No. 8 stopes are arranged along the ore body strike. Four sections are arranged in the panel, which are connected with the sectional drift and the stope respectively through the stope connection path. Each section serves 3-4 layers, with a mining height of 2.5-3 m, a controlled roof height of 4 m and a backfilling height of 2.5 m. For the point-pillar upward layering and backfilling mining method, the panel is divided into three sections in the horizontal direction, with 5 m continuous pillars between the sections, and the size of point-pillars in each section is (4-5) m × (4-5) m. Vertical direction is divided into 4 sections, each section serves 4-5 layers, and the mining height of each layer is 3-3.5 m. Out-in-vein mining method, the out-in-vein mine horizontal roadway is driven at certain height in the height direction, arranging out-in-vein mine roadway at the foot wall of the ore body, driving out-in-vein mine roadway to the ore body every 36 m, and then driving, cutting and ventilating up the mountain in the center of the ore body.

The advantages, disadvantages and main technical and economic indexes of drift upward layering and backfilling mining method (Method 1), room-pillar alternating upward layering and backfilling mining method (Method 2) and point-pillar upward layering and backfilling mining method (Method 3) are shown in Table 6-1.

6 Mining Optimization and Stability Analysis of Stopes

Comprehensive comparison of three mining methods　　　　Table 6-1

Method	Advantage	Disadvantage	Main technical and economic indexes
Method 1	① It is conducive to ground pressure control. ② Mining ore in the drift is conducive to safe mining. ③ The exposed surface of mining is small, which is beneficial to realize low sedimentation of mining	① Drift top tight is difficult. ② Stope production capacity and mining efficiency is low. ③ One-step drift must be cemented and filled, which increases the backfilling cost	Stope production capacity: 73.4 t/d; Mining efficiency: 13.6 t/shift; Ore loss rate: 9%; Ore dilution rate: 5%; Stripping ratio: 24 m^3/kt; Mining cost: 67.96 yuan/t
Method 2	①Production capacity is high. ②Mining efficiency is high. ③Blasting cost is low. ④Mining safety is large	① Alternating rising mining process is complex, management is difficult. ② Quantities of mining and cutting is large	Panel production capacity: 527.59 t/d; Mining efficiency: 57.09 t/shift; Ore dilution rate: 6%; Ore loss rate: 9%; Stripping ratio: 28.79 m/kt; Mining cost: 53.09 yuan/t
Method 3	①Stripping ratio is relatively small. ②Production capacity is relatively high. ③ Using zoning method reduces mining exposed area, which is conducive to reducing mining sedimentation	① Intermediate stope is cemented backfilled. ②Production management difficulty is great	Stope production capacity: 103.5 t/d; Mining efficiency: 17.4 t/shift; Ore loss rate: 16.2%; Ore dilution rate: 6%; Stripping ratio: 48.7 m^3/kt; Mining direct cost: 61.61 yuan/t

Through comprehensive comparison, it is found that the mining dilution rate and loss rate of the room-pillar alternating upward backfilling mining method are relatively low, the mining safety is very good, and it has good adaptability to deep high stress and high rock burst proneness. At the same time, it can greatly improve the production capacity in the deep mining of similar ore bodies, meet the production capacity requirements put forward by mines, and is suitable for wide application in suitable areas of deep mining.

6.2 Optimization of structure parameters in deep stope

The mining dilution rate and loss rate of the room-pillar alternating upward layering and backfilling mining method are relatively low, and the mining safety is very good, which has good adaptability to deep high stress and high rock burst proneness. The mining of Sanshandao deep deposit adopting this method can well meet the requirements of production capacity and safety. At present, the structure parameters of mines adopting this mining method in China are different in size, and the width of rooms and pillars ranges from 6 m to 15 m. In view of the deep high stress, high rock burst proneness and the increasing mining scale and difficulty of Sanshandao Gold Mine, the method of simply relying on engineering experience analogy is no longer applicable. The optimization of stope structure parameters through numerical simulation analysis is of great significance for realizing large-capacity and high-efficiency economic mining and meeting production safety requirements.

6.2.1 Weighted multiple weighting optimization model

(1) Concept definition

1) u is the alternative evaluation index set:
$$u = \{u_1, u_2, u_3, \cdots, u_9\} \quad (6\text{-}1)$$
where u_i—each evaluation index, $i = 1, 2, 3, \cdots, 9$.

2) V is the alternative set:
$$V = \{V_1, V_2, V_3\} \quad (6\text{-}2)$$
where V_j—each scheme, and is a fuzzy subset on u, $j = 1, 2, 3$.

(2) Quantitative index membership degree matrix

The quantitative index membership degree matrix is determined by the membership degree function method, and the quantitative index is constructed into the target eigenvalue matrix:

$$Y = \begin{bmatrix} y_{11} & y_{12} & \cdots & y_{1n} \\ y_{21} & y_{22} & \cdots & y_{2n} \\ \cdots & \cdots & \cdots & \cdots \\ y_{m1} & y_{m2} & \cdots & y_{mn} \end{bmatrix} \quad (6\text{-}3)$$

where n—scheme number, $n = 1, 2, 3$;

m—quantitative index number, $m = 1, 2, 3, 4, 5, 6$.

The relative membership degree of quantitative indexes is normalized by the following criteria:

For the larger the value, the better the index, use

$$r_{ij} = \frac{y_{ij}}{\max y_{ij}} \tag{6-4}$$

to make normalization, where r_{ij}—the factor of relative optimization membership degree matrix.

For the smaller the value, the better the index, use

$$r_{ij} = \frac{\min y_{ij}}{y_{ij}} \tag{6-5}$$

to make normalization.

Through normalization, the target optimization membership degree matrix is obtained:

$$R_1 = \begin{bmatrix} r_{11} & r_{12} & \cdots & r_{1n} \\ r_{21} & r_{22} & \cdots & r_{2n} \\ \cdots & \cdots & \cdots & \cdots \\ r_{m1} & r_{m2} & \cdots & r_{mn} \end{bmatrix} \tag{6-6}$$

(3) Relative membership degree matrix of non-quantitative indexes

The relative membership degree of non-quantitative indexes is normalized by the relative binary comparison method, and the ranking scale is determined according to the following principles:

If R_k is more important than R_l, make $r_{kl} = 1$, $r_{lk} = 0$;

If R_k is as important as R_l, make $r_{kl} = 0.5$, $r_{lk} = 0.5$;

If R_l is more important than R_k, make $r_{kl} = 0$, $r_{lk} = 1$;

where $k = 1, 2, \cdots, m$, $l = 1, 2, \cdots, n$.

According to the above scale, a non-quantitative eigenvector matrix is established:

$$R_2 = \begin{bmatrix} r_{11} & r_{12} & \cdots & r_{1n} \\ r_{21} & r_{22} & \cdots & r_{2n} \\ \cdots & \cdots & \cdots & \cdots \\ r_{m1} & r_{m2} & \cdots & r_{mn} \end{bmatrix} \tag{6-7}$$

The sum of each row of the non-quantitative eigenvector matrix is sorted by descending order, and the importance of each index is sorted. If the value is the same, the order is the same. It is sorted by the importance, according to the mood operator and quantitative scale table shown in Table 6-2, the relative membership degree matrix of non-quantitative indexes is established.

Relative membership degree relationship table between mood operator and quantitative scale

Table 6-2

Mood operator	Same	Slightly	A little	Comparatively	Obviously	Remarkably	Very	Extraordinarily	Extremely	Utmost	Incomparably
Quantitative	0.50	0.55	0.60	0.65	0.70	0.75	0.80	0.85	0.90	0.95	1
Scale	0.525	0.575	0.625	0.675	0.725	0.775	0.825	0.875	0.925	0.975	
Relative	1.0	0.818	0.667	0.538	0.429	0.333	0.25	0.176	0.111	0.053	0
Membership degree	0.905	0.739	0.60	0.481	0.379	0.29	0.212	0.143	0.081	0.026	

(4) Comprehensive membership degree matrix

Combining quantitative index membership degree matrix with non-quantitative index membership degree matrix, the comprehensive membership degree matrix of optimization index is established.

(5) Establishing multiple weighting weight vectors based on weighted judgment

The variation coefficient method, fuzzy criterion method and entropy method are used to calculate the weight value of each optimization index, and the optimization weight vector is established. The consistency of the relative importance order of the optimization indexes determined by each weighting method is evaluated. On this basis, the weighted average method is used to calculate the average weight of the three weighting schemes as the final combined weight vector.

(6) The comprehensive membership degree matrix and combined weight vector of quantitative and non-quantitative indexes are combined, the membership degree of each scheme is calculated, and the best scheme is determined.

6.2.2 Membership degree of optimization index based on numerical simulation

With the development of computer technology, three-dimensional finite element analysis software represented by FLAC3D has been widely and deeply applied in mines, which can be used to simulate stope stability and stress and displacement during mining, and provide reliable quantitative data and basis for the design and construction of mining process.

In order to optimize the layout of stope width of one-step mining and two-step mining in the process of alternating upward layering and backfilling of deep room-pillar in Sanshandao Gold Mine, five sets of matching schemes of one-step mining width and two-step mining width are designed by engineering analogy, as shown in Table 6-3, FLAC3D models are established for each scheme to simulate mining, and weighted multiple weighting optimization models of stope structure parameters are constructed. The middle wall roof sedimentation (final excavation boundary roof), the vertical displacement of hanging wall, the vertical displacement in the stope, X (along the ore body direction) direction displacement, X direction stress, Y (vertical strike of ore body) direction displacement and Y direction stress are taken as the quantitative indexes for optimization, and the optimization is carried out with reference to the non-quantitative indexes such as comprehensive efficiency, relative production capacity, equipment matching degree and construction difficulty.

Width matching proposed scheme of one-step mining and two-step mining

Table 6-3

	Scheme 1	Scheme 2	Scheme 3	Scheme 4	Scheme 5
One-step mining width(m)	8	10	10	12	12
Two-step mining width(m)	8	8	10	10	12

The room pillar alternating upward layering and backfilling mining method is adopted, and the mechanized panel layout is adopted, and the adjacent one-step mining room and two-step mining room are alternately mined. Numerical simulation continuously arranges eight mining rooms in the panel. According to the alternation of excavation and backfilling, each scheme is divided into six steps for excavating, and the mining and backfilling state of each step is as follows:

Step 1: The mining height of stopes 2, 4, 6 and 8 is 9.5 m, by top tight filling, and the mining height of stopes 1, 3, 5 and 7 is 0;

Step 2: The mining height of stopes 2, 4, 6 and 8 is 9.5 m, by top tight filling, and the mining height of stopes 1, 3, 5 and 7 is 17.5 m, by top tight filling;

Step 3: The mining height of stopes 2, 4, 6 and 8 is 27 m, by top tight filling, and the mining height of stopes 1, 3, 5 and 7 is 17.5 m, by top tight filling;

Step 4: The mining height of stopes 2, 4, 6 and 8 is 27 m, by top tight filling, and the mining height of stopes 1, 3, 5 and 7 is 35 m, by top tight filling;

Step 5: The mining height of stopes 2, 4, 6 and 8 is 42 m, by top tight filling, and the mining height of stopes 1, 3, 5 and 7 is 35 m, by top tight filling;

Step 6: The mining height of stopes 2, 4, 6 and 8 is 42 m, by top tight filling, and the mining height of stopes 1, 3, 5 and 7 is 42 m, by top tight filling.

In the numerical simulation, the middle section in -555 m is taken as the simulation object, and the mechanized panel mining is divided into 8 continuous stopes along the ore body strike, which are arranged vertically along the ore body strike, and the actual buried depth of the panel is form -510 m to -555 m. The size of FLAC3D numerical model is 400 m in X (along the ore body strike direction), 300 m in Y (the vertical ore body strike direction) and 200 m in Z (vertical direction). The actual buried depth of the model is -430 m at the top and -630 m at the bottom. The ore body thickness is 30 m, the dip angle is 45°, and the panel height is 45 m. The FLAC3D numerical calculation model is shown in Figure 6-4, which includes five media: ore body, backfilling body, hanging wall, foot wall and fault. The mining sequence of one-step mining and two-step mining is shown in Figure 6-5.

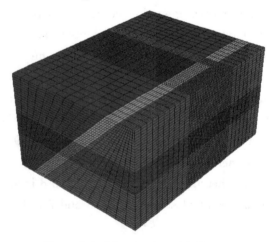

Figure 6-4 Numerical model diagram

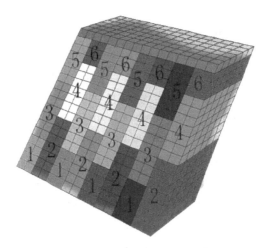

Figure 6-5 Schematic diagram of one-step mining and two-step mining sequence

According to the physical and mechanical test results of deep ore and rock in Sanshandao Gold Mine, referring to *Standard for Engineering Classification of Rock Mass* GB/T 50218—2014 and *Code for Investigation Geotechnical Engineering* GB 50021—2001 and the engineering geological conditions of deep mining in Sanshandao Gold Mine, the rock mass mechanical parameters of hanging wall surrounding rock, foot wall surrounding rock, ore body, backfilling body and fault in the numerical calculation model are comprehensively selected and shown in Table 6-4.

Mechanical parameters of rock mass   Table 6-4

Position	Bulk modulus (GPa)	Shear modulus (GPa)	Cohesion (MPa)	Internal friction angle(°)	Tensile strength (MPa)
Hanging wall	3.37	2.8	11.4	31	4.37
Foot wall	5.48	3.45	42.8	37	5.98
Fault	0.595	3.78	0.128	18	0.0198
Ore body	2.51	1.35	21.5	33	3.44

In the process of optimization simulation calculation of structure parameters of deep stope, the buried depth of stope is below -500 m, so according to the results of in-situ stress measurement and the occurrence state of deep structure of Sanshandao Gold Mine, the measured in-situ stress is applied to the boundary of the model according to gradient, and the bottom of the model limits the displacement in Z direction, while the two surfaces perpendicular to X axis limit the displacement in X direction and the two surfaces perpendicular to Y axis limit the displacement in Z direction.

According to the design of stope structure parameter optimization scheme, five stope structure parameter schemes were simulated by FLAC3D numerical simulation software, and seven indexes of maximum roof sedimentation, maximum vertical displacement of hanging wall, maximum vertical displacement of stope, maximum X-direction displacement of stope, maximum X-direction stress of stope, maximum Y-direction displacement of stope and maximum Y-direction stress of each scheme were extracted by FISH language embedded in FLAC3D as quantitative indexes for weighted multiple weighting optimization of stope structure parameters (see Table 6-5).

Numerical simulation results of five schemes   Table 6-5

Evaluating index	Scheme 1	Scheme 2	Scheme 3	Scheme 4	Scheme 5
Roof sedimentation(m)	0.1144	0.1398	0.1421	0.1621	0.1650
Maximum vertical displacement of stope(m)	0.2394	0.2671	0.2675	0.2759	0.2784
Maximum vertical displacement of hanging wall(m)	0.1232	0.1398	0.1459	0.162	0.1650
Maximum X-direction displacement of stope(m)	0.1562	0.1732	0.1873	0.1012	0.1055
Maximum X-direction stress of stope(MPa)	58.981	65.491	61.982	73.78	63.394
Maximum Y-direction displacement of stope(m)	0.0943	0.0991	0.0937	0.1022	0.1128
Maximum Y-direction stress of stope(MPa)	33.169	38.231	39.603	42.13	44.926

According to the occurrence state of deep ore body, fracture structure distribution, blasting and mining and transportation equipment in Sanshandao Gold Mine, combined with field production data investigation and experience analysis, the comprehensive efficiency, relative production capacity, equipment matching degree and construction difficulty of each stope structure scheme are selected as the optimization non-quantitative indexes, and the quantitative indexes of numerical simulation are integrated. The optimization index pairs of deep stope structure parameters are shown in Table 6-6.

6 Mining Optimization and Stability Analysis of Stopes

Comparison of optimization evaluation indexes of stope structure parameters Table 6-6

Evaluating index	Scheme 1 Index	Scheme 1 Sequence	Scheme 2 Index	Scheme 2 Sequence	Scheme 3 Index	Scheme 3 Sequence	Scheme 4 Index	Scheme 4 Sequence	Scheme 5 Index	Scheme 5 Sequence
Roof sedimentation(m)	0.1144	1	0.1398	2	0.1421	3	0.1621	4	0.1650	5
Maximum vertical displacement of stope(m)	0.2394	1	0.2671	2	0.2675	3	0.2759	4	0.2784	5
Maximum vertical displacement of hanging wall(m)	0.1232	1	0.1398	2	0.1459	3	0.162	4	0.1650	5
Maximum X-direction displacement of stope(m)	0.1562	1	0.1732	2	0.1873	3	0.1012	4	0.1055	5
Maximum X-direction stress of stope(MPa)	58.981	1	65.491	4	61.982	2	73.78	5	63.394	3
Maximum Y-direction displacement of stope(m)	0.0943	2	0.0991	3	0.0937	1	0.1022	4	0.1128	5
Maximum Y-direction stress of stope(MPa)	33.169	1	38.231	2	39.603	3	42.13	4	44.926	5
Construction difficulty	More difficult	3	Easier	2	Easy	1	Easy	1	Difficult	4
Equipment matching degree	Worse	4	Better	3	Good	1	Second best	2	Poor	5
Relative production capacity	Poor	4	Worse	3	Better	2	Better	2	Good	1
Comprehensive efficiency	Poor	5	Worse	4	Second best	2	Good	1	Better	3

According to Table 6-6, the characteristic matrix of quantitative evaluation index is:

$$Y_{1\text{-}7} = \begin{bmatrix} 0.1144 & 0.1398 & 0.1421 & 0.1621 & 0.1650 \\ 0.2394 & 0.2671 & 0.2675 & 0.2759 & 0.2784 \\ 0.1232 & 0.1398 & 0.1459 & 0.1620 & 0.1650 \\ 0.1562 & 0.1732 & 0.1873 & 0.1012 & 0.1055 \\ 58.981 & 65.491 & 61.982 & 73.78 & 63.394 \\ 0.0943 & 0.0991 & 0.0937 & 0.1022 & 0.1128 \\ 33.169 & 38.231 & 39.603 & 42.13 & 44.926 \end{bmatrix}$$

The relative membership degree matrix is:

$$Y_{1\text{-}7} = \begin{bmatrix} 1 & 0.82 & 0.81 & 0.71 & 0.69 \\ 1 & 0.90 & 0.89 & 0.87 & 0.86 \\ 1 & 0.88 & 0.84 & 0.76 & 0.75 \\ 1 & 0.94 & 0.89 & 0.85 & 0.84 \\ 1 & 0.90 & 0.95 & 0.80 & 0.93 \\ 0.99 & 0.95 & 1 & 0.92 & 0.83 \\ 1 & 0.87 & 0.84 & 0.79 & 0.74 \end{bmatrix}$$

The eigenvector matrix of non-quantitative index construction difficulty is:

$$Y_8 = \begin{bmatrix} 0.5 & 0 & 0 & 0 & 1 \\ 1 & 0.5 & 0 & 0 & 1 \\ 1 & 1 & 0.5 & 0.5 & 1 \\ 1 & 1 & 0.5 & 0.5 & 1 \\ 0 & 0 & 0 & 0 & 0.5 \end{bmatrix} \begin{bmatrix} 3 \\ 2 \\ 1 \\ 1 \\ 4 \end{bmatrix}$$

The eigenvector matrix of equipment matching degree is:

$$Y_9 = \begin{bmatrix} 0.5 & 0 & 0 & 0 & 1 \\ 1 & 0.5 & 0 & 0 & 1 \\ 1 & 1 & 0.5 & 1 & 1 \\ 1 & 1 & 0 & 0.5 & 1 \\ 0 & 0 & 0 & 0 & 0.5 \end{bmatrix} \begin{bmatrix} 4 \\ 3 \\ 1 \\ 2 \\ 5 \end{bmatrix}$$

The eigenvector matrix of relative production capacity is:

$$Y_{10} = \begin{bmatrix} 0.5 & 0 & 0 & 0 & 0 \\ 1 & 0.5 & 0 & 0 & 0 \\ 1 & 1 & 0.5 & 0.5 & 0 \\ 1 & 1 & 0.5 & 0.5 & 0 \\ 1 & 1 & 1 & 1 & 0.5 \end{bmatrix} \begin{bmatrix} 4 \\ 3 \\ 2 \\ 2 \\ 1 \end{bmatrix}$$

6 Mining Optimization and Stability Analysis of Stopes

The eigenvector matrix of comprehensive efficiency is:

$$Y_{11} = \begin{bmatrix} 0.5 & 0 & 0 & 0 & 0 \\ 1 & 0.5 & 0 & 0 & 0 \\ 1 & 1 & 0.5 & 0 & 1 \\ 1 & 1 & 0.5 & 1 & 1 \\ 1 & 1 & 0 & 0 & 0.5 \end{bmatrix} \begin{bmatrix} 5 \\ 4 \\ 2 \\ 1 \\ 3 \end{bmatrix}$$

The relative membership degree vector of each non-quantitative evaluation index is:

$$R_8 = [0.379 \quad 0.739 \quad 1 \quad 1 \quad 0.176]$$
$$R_9 = [0.429 \quad 0.739 \quad 1 \quad 0.818 \quad 0.143]$$
$$R_{10} = [0.333 \quad 0.538 \quad 0.905 \quad 0.905 \quad 1]$$
$$R_{11} = [0.429 \quad 0.481 \quad 0.818 \quad 1 \quad 0.739]$$

To sum up, the comprehensive membership degree matrix of all quantitative indexes and non-quantitative indexes is:

$$R = \begin{bmatrix} 1 & 0.82 & 0.81 & 0.71 & 0.69 \\ 1 & 0.90 & 0.89 & 0.87 & 0.86 \\ 1 & 0.88 & 0.84 & 0.76 & 0.75 \\ 1 & 0.94 & 0.89 & 0.85 & 0.84 \\ 1 & 0.90 & 0.95 & 0.80 & 0.93 \\ 0.99 & 0.95 & 1 & 0.92 & 0.83 \\ 1 & 0.87 & 0.84 & 0.79 & 0.74 \\ 0.379 & 0.739 & 1 & 1 & 0.176 \\ 0.429 & 0.739 & 1 & 0.818 & 0.143 \\ 0.333 & 0.538 & 0.905 & 0.905 & 1 \\ 0.429 & 0.481 & 0.818 & 1 & 0.739 \end{bmatrix}$$

6.2.3 Multiple weighting optimization based on weighted discrimination

(1) Coefficient of variation method

The weighting principle of coefficient of variation method is to maximize the weight dispersion of each index, and to give smaller weight to the index with small numerical difference and larger weight to the index with large numerical difference.

According to the above comprehensive membership degree matrix, the mean and variance of each row vector of matrix are calculated:

$$\mu_i = \frac{1}{n}\sum_{j=1}^{n} r_{ij} \tag{6-8}$$

$$s_i = \sqrt{\frac{\sum_{j=1}^{n}(r_{ij}-\mu)^2}{n-1}} \qquad (6-9)$$

The coefficient of variation of each index is calculated by the following equation:

$$\beta_i = \frac{s_i}{\mu_i} \qquad (6-10)$$

The obtained coefficient of variation matrix is:
$\omega_1 = $ (0.096519, 0.08607, 0.08318, 0.08969, 0.099231, 0.109658, 0.081274, 0.38966, 0.399158, 0.255131, 0.245602)^T

The coefficient of variation matrix isnormalized, the weights of each index weighted by the coefficient of variation method are as follows:
$\omega_1 = $ (0.049876, 0.44476, 0.042983, 0.046347, 0.051277, 0.056665, 0.041998, 0.201356, 0.206264, 0.131838, 0.126914)^T

(2) Weighting by fuzzy criterion

According to the comprehensive membership degree matrix, the weight of each index is calculated according to equation (6-11):

$$\omega_i = \frac{2\sum_{j=1}^{n}1-r_{ij}}{n(n-1)} \qquad (6-11)$$

The weight matrix of fuzzy criterion is obtained as follows:
$\omega_2 = $ (0.097, 0.048, 0.077, 0.048, 0.042, 0.031, 0.076, 0.1706, 0.1871, 0.1319, 0.1533)^T

The weight matrix is normalized, the weights of each index weighted by the fuzzy criterion method are as follows:
$\omega_2 = $ (0.091345, 0.045201, 0.072511, 0.045201, 0.039551, 0.029192, 0.071569, 0.160655, 0.176194, 0.134211, 0.144363)^T

(3) Weighting by entropy method

Entropy method is based on disorder degree, and its weighting principle is that the greater the value, the higher the disorder degree and the smaller the weighting. On the contrary, the smaller the value, the smaller the disorder degree and the greater the weight.

The entropy of each index is calculated by using equation (6-12):

$$h_j = -(\ln n)^{-1}\sum_{i=1}^{m}k_{ij}\ln k_{ij} \qquad (6-12)$$

where m—scheme number;

n—index number.

After obtaining the entropy value of each index, the weight value of each index by is calculated by equation (6-13):

$$\omega_i = \frac{1-h_i}{\sum_{j=1}^{n}(1-h_j)} \quad (6\text{-}13)$$

The weight matrix of entropy method is obtained as follows:

$\omega_3 =$ (0.546313, 0.513945, 0.532746, 0.514095, 0.510512, 0.504257, 0.532101, 0.623797, 0.634735, 0.580469, 0.59218)^T

The weight matrix is normalized, the weights of each index weighted by entropy method are as follows:

$\omega_3 =$ (0.089778, 0.084459, 0.087549, 0.084484, 0.083895, 0.082867, 0.087443, 0.102511, 0.1043090.095391, 0.097316)^T

The index weights determined by each weighting scheme are sorted according to the ascending order, and the sorting numbers are integers from 1 to 11. After sorting, the combined sorting vector of three weighting schemes can be obtained:

$$\begin{bmatrix} 7 & 3 & 6 & 4 & 2 & 1 & 5 & 10 & 11 & 8 & 9 \\ 7 & 3 & 6 & 4 & 2 & 1 & 5 & 10 & 11 & 8 & 9 \\ 7 & 3 & 6 & 4 & 2 & 1 & 5 & 10 & 11 & 8 & 9 \end{bmatrix}$$

It can be seen that the relative importance of each index weight determined by the three weighting methods is completely consistent. Therefore, in order to reduce the contingency of a single weighting method, the weighted average method can be used to increase the reliability of weighting, and then the optimization weight vector of optimization index can be obtained.

The weight vector of each optimization index obtained after weighted average calculation is as follows:

$\omega_4 =$ (0.081603, 0.063535, 0.072529, 0.063552, 0.062692, 0.062501, 0.072125, 0.12482, 0.1351480.100948, 0.113433)^T

The weighted combination weight vector and the comprehensive membership degree matrix of optimization index is combined, the membership degrees of five stope structure parameter colposition schemes are calculated as follows:

$R =$ [0.665477 0.727196 0.866888 0.827150 0.608348]

Through multiple weighting optimization, it can be seen that the optimization order of the five schemes is 3, 4, 2, 1, 5. The optimization results show that the stope width of one-step mining and two-step mining in Sanshandao Gold

Mine is 10 m, and the stope length is the thickness of ore body.

6.3 Spatio-temporal variation of displacement and stress in stope

6.3.1 Numerical model for the mining stope

The fluid-structure coupling model of deep mining is established by FLAC3D numerical simulation software. According to the occurrence depth of deep ore body in Sanshandao Gold Mine, the elevation of the model is from -1180 m to -380 m. The coordinate system takes the ore body strike as the X axis, the ore body thickness direction as the Y axis and the vertical direction as the Z axis. The length of the model in X direction is 400 m, the length in Y direction is 800 m and the height in Z direction is 800 m, which is divided into 239508 units and 256878 nodes. According to the engineering geological and hydrological conditions on site, the model is divided into four groups, namely the hanging wall rock group, foot wall rock group, fault rock group and ore body group, as shown in Figure 6-6.

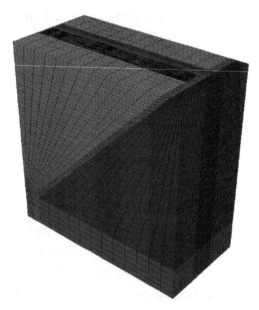

Figure 6-6 Fluid-structure coupling model for deep mining

(1) Mechanical parameters

The mechanical parameters and permeability coefficients of each medium in the fluid-structure coupling analysis model are shown in Table 6-7, in which the permeability coefficient is obtained by geometric calculation and Monte-carlo structure simulation correction based on the investigation and statistics of joints and fissures.

Mechanical parameters of fluid-structure coupling simulation　　　Table 6-7

Position	Bulk modulus (GPa)	Shear modulus (GPa)	Cohesion (MPa)	Internal friction angle(°)	Tensile strength (MPa)	Permeability coefficient (m/s)
Hanging	3.37	2.8	11.4	31	4.37	6.2×10^{-7}
Foot wall	5.48	3.45	42.8	37	5.98	6.2×10^{-7}
Fault	0.60	0.38	0.13	18	0.02	1.4×10^{-5}
Ore body	2.51	1.35	21.5	33	3.44	8.3×10^{-7}

(2) Boundary conditions

The maximum principal stress, minimum principal stress and vertical principal stress are calculated according to the deep in-situ stress field model, and the gradient stress boundary is applied to the model, and the gravity field is applied to the whole model.

(3) Simulation scheme

The -555 m middle section mining is simulated by using the room-pillar alternating upward layering and backfilling mining method, and the disturbance of mining to stope is studied, and the displacement, stress and plastic failure of stope and its top pillar are analyzed. The excavation steps and sequence are shown in Figure 6-7.

Studying the influence of mining on stope stress and displacement, and summarizing the variation law of stress and displacement in the whole panel during alternating mining have guiding significance for realizing safe mining and targeted stope support. The ore body on the north side of F3 in the middle section in -555 m is simulated by the upward layering and backfilling in the panel with the panel length of 80 m. According to the optimization results of stope structure parameters, each stope has the width of 10 m and the thickness of the ore body is the stope length. Eight stopes are arranged along the ore body strike in the panel for the room-pillar alternately upward mining.

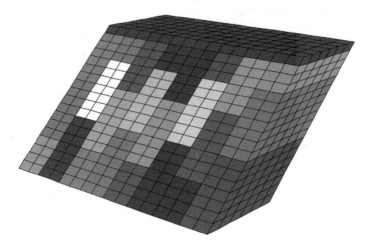

Figure 6-7 Schematic diagram of middle section mining

6.3.2 Displacement field analysis

In the process of stope mining, because of the existence of gob or the weak strength of backfilling body, the surrounding rock will be displaced towards stope. The sedimentation of stope roof to a certain extent will cause small-scale roof caving, and even large-scale roof caving in extreme cases.

It can be seen from Figure 6-8 that the maximum X-direction displacement of stope caused by mining changes with the mining process as follows: the maximum X-direction displacements of ore body and backfilling body in panel stope in Step 1 are −3.663 cm and 3.440 cm; the maximum X-direction displacements of ore body and backfilling body in panel stope in Step 2 are −4.720 cm and 6.682 cm; the maximum X-direction displacements of ore body and backfilling body in panel stope in Step 3 are −7.236 cm and 6.414 cm; the maximum X-direction displacements of ore body and backfilling body in panel stope in Step 4 are −8.816 cm and 9.021 cm; the maximum X-direction displacements of ore body and backfilling body in panel stope in Step 5 are −11.015 cm and 8.515 cm; the maximum X-direction displacements of ore body and backfilling body in panel stope in Step 6 are −10.400 cm and 10.007 cm. It can be seen from the data of the maximum X-direction displacement that the X-direction displacement of panel is compressed and concentrated towards the middle of panel in each mining step due to the weak strength of backfilling body and surrounding rock compression. It can be found in the comprehensive analysis of the X-direction displacement nephogram of the middle section of back panel after each mining step that the X-direction displacement of No. 1 to No. 4 stope is positive, No. 5 stope is the boundary zone of positive and negative X-direction displacement, and No. 6 to No. 8 stope is negative. Comparing the two areas with opposite displacement in X direction, the positive displacement area is obviously larger than the negative displacement area. Because the X axis of the rocks in No. 1 to No. 8 stope is distributed in sequence, this phenomenon can be explained from the geological structure. F3 fault with a width of about 25 m exists in the one wing of No. 8 stope, and the existence of F3 fault causes the stress release of surrounding rock mass, which is less than the stress of surrounding rock in the one wing of No. 1 stope with the same depth, which leads to the positive displacement area being larger than the negative displacement area after panel mining.

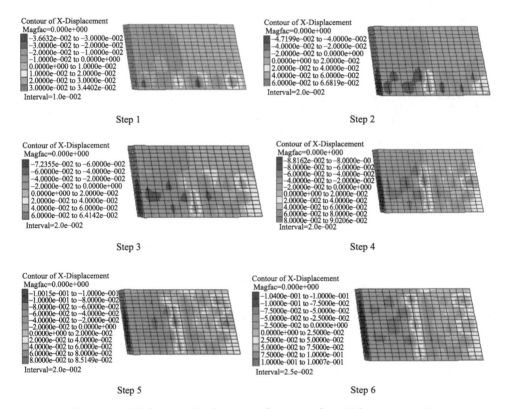

Figure 6-8　X-direction displacement of stope in the middle section mining

As the ore body dip angle is nearly 45°, the displacement of the hanging wall rock mass towards the stope has an important influence on the stope safety. The Y-direction displacement of the hanging wall rock mass in the stope at each step of the middle section mining is shown in Figure 6-9. The Y-direction displacement of hanging wall caused by mining in step 1 is 1.2446 cm; The Y-direction displacement of hanging wall caused by mining in step 2 is 3.8950 cm; The Y-direction displacement of hanging wall caused by mining in step 3 is 6.3405 cm. With the enlargement of the exposed area of hanging wall, the Y-direction displacement toward the stope also gradually increases. After the introduction of the middle section mining, the maximum Y-direction displacement reaches the peak value of 6.3405 cm.

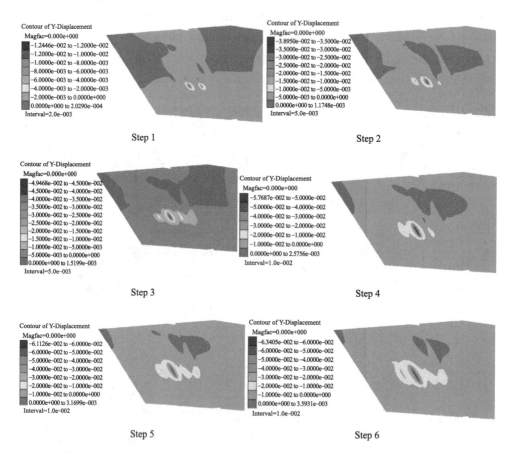

Figure 6-9 Y-direction displacement of hanging wall in middle section mining

6.3.3 Analysis of stress field and plastic failure

Stress redistribution is caused by stope mining, and stress state is the most important factor affecting stope stability. In order to realize safe mining, it is necessary to analyze the stress distribution characteristics of stope, identify the potential damage areas caused by stress concentration, tensile stress and shear stress, and focus on strengthening the support in the production process. The maximum principal stress, minimum principal stress and plastic failure in the middle section in —555 m are shown in Figure 6-10 and Figure 6-11, in which the tensile stress is positive and the compressive stress is negative, and the maximum principal stress and minimum principal stress are determined by the absolute value of stress.

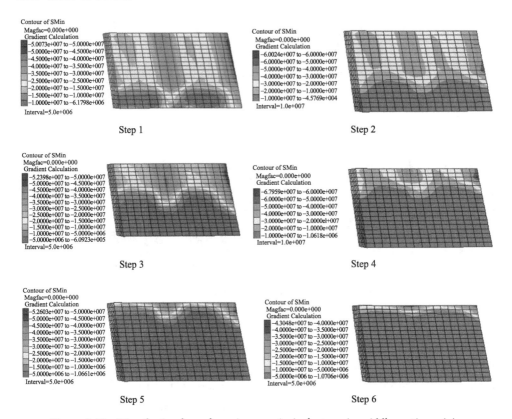

Figure 6-10 Distribution law of maximum principal stress in middle section mining

It can be seen from the maximum principal stress nephogram in Figure 6-10 that the maximum principal stress values of the six mining steps are 50.073 MPa, 60.024 MPa, 52.398 MPa, 67.959 MPa, 52.603 MPa and 43.048 MPa respectively. It can be seen from this analysis that before the whole panel mining is finished, the maximum principal stress in the mining process of stopes 2, 4, 6 and 8 is greater than that in the mining process of stopes 1, 3, 5 and 7, indicating that the stress concentration caused by two-step mining is greater than that

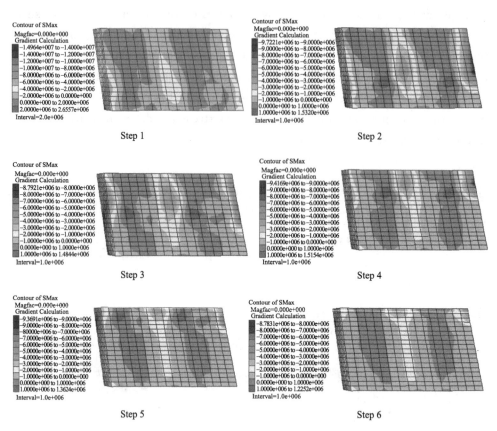

Figure 6-11　Distribution law of minimum principal stress in middle section mining

caused by one-step mining. The maximum principal stress in the middle and two wings is high, and the overall nephogram shows an obvious M shape. The extreme value of the maximum principal stress mainly appears in the surrounding rock of No. 1 stope, and the stress concentration on the roof of No. 5 stope is also obvious. At the same time, the corner of the contact surface between the backfilling body and the ore body in each step of mining is also a stress concentration area, so the support shall be strengthened when the mining operation reaches the above three positions.

It can be seen from the minimum principal stress nephogram in Figure 6-11 that the minimum principal stress of each mining step is 2.656 MPa, 1.532 MPa, 1.484MPa, 1.515 MPa, 1.362 MPa and 1.225 MPa respectively. When the mining height of adjacent stopes on both sides exceeds that of the current stope, the strength of backfilling bodies on both sides is small and the internal horizontal stress is low, which leads to obvious tensile stress of ore bodies in the current stope. Its distribution area has a certain rule, that is, the one-step mining stage mainly exists in the roof of No. 2 stope and No. 6 stope, and the two-step mining stage mainly exists in the roof of No. 3 stope and No. 7 stope.

Figure 6-12 shows that the plastic zone increases with the mining. Among them, the shear failure mainly occurs in the backfilling body, while the tensile failure corresponds to the tensile stress distribution, and occurs in the area near the hanging wall rock of the inverted ore body. Therefore, in actual mining, when each layered mining approaches the confining pressure of hanging wall, attention shall be paid to the surrounding rock spalling of hanging wall rock mass.

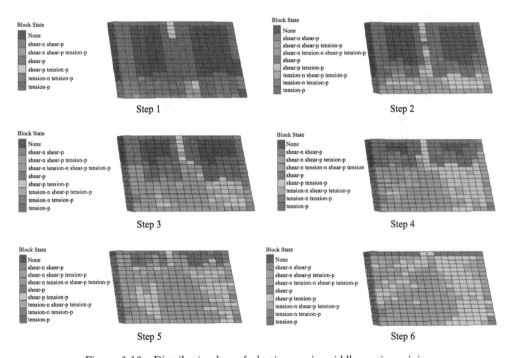

Figure 6-12　Distribution law of plastic zone in middle section mining

6.4　Spatio-temporal variation of displacement and stress in multiple mining section mining

6.4.1　Simulation scheme

The model, mechanical parameters and boundary conditions used in the simulation of deep multiple middle section mining are the same as those in -555 m middle section mining. The simulation scheme is shown in Figure 6-13, which simulates the mining of a 45 m middle section of deep ore body, and simulates and analyzes the displacement, stress and plastic failure of each middle section of deep mining in the future.

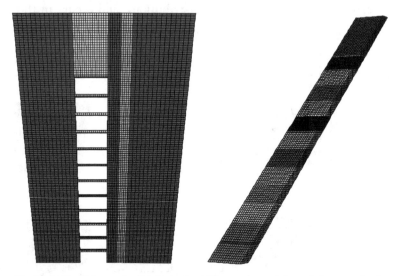

Figure 6-13　Schematic diagram of middle section mining of deep ore body

6.4.2　Displacement field analysis

It can be seen from Figure 6-14 and Figure 6-15 that with the increase of mining depth, the horizontal displacement of mining increases continuously, and the maximum horizontal displacement increases from 6.24 cm to 7.32 cm. With the increase of mining depth, the vertical displacement caused by mining also increases, and the maximum value of vertical displacement increases from 6.78 cm to 8.17 cm.

6.4.3　Analysis of stress field and plastic failure

Figure 6-16 and Figure 6-17 show the maximum principal stress nephogram and minimum principal stress nephogram after simulated mining in deeper horizontal multiple middle sections. The simulation results show that with the increase of mining depth, both the maximum principal stress and the minimum principal stress increase. The maximum and minimum principal stress extremum of simulated mining in each deeper middle section is shown in Table 6-8.

■ 6 Mining Optimization and Stability Analysis of Stopes ■

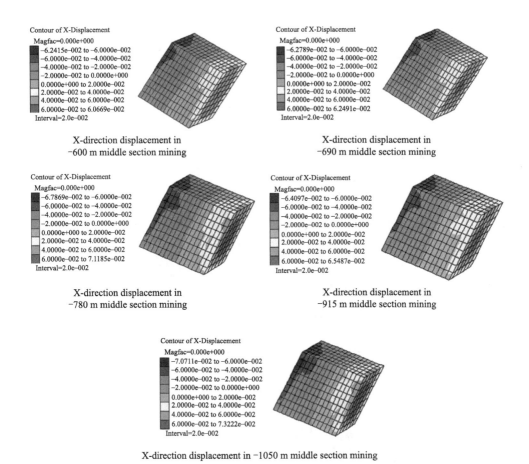

Figure 6-14 X-direction displacement diagram of multiple middle section mining

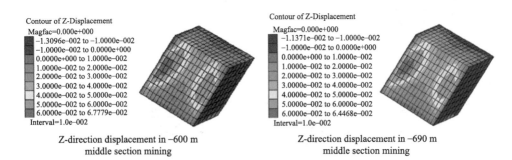

Figure 6-15 Z-direction displacement diagram of multiple middle section mining (one)

177

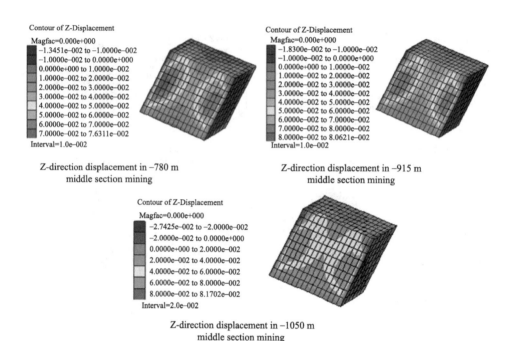

Figure 6-15 Z-direction displacement diagram of multiple middle section mining (two)

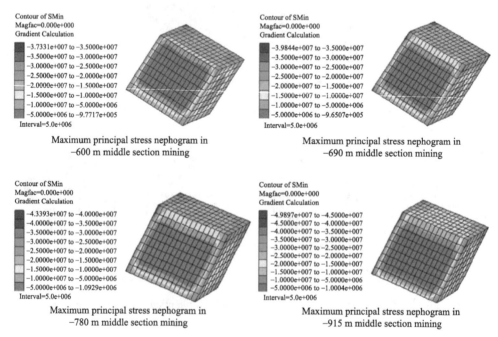

Figure 6-16 Distribution law of maximum principal stress in deep multiple middle section mining (one)

■ 6 Mining Optimization and Stability Analysis of Stopes ■

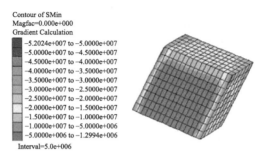

Maximum principal stress nephogram in −1050 m middle section mining

Figure 6-16 Distribution law of maximum principal stress in deep multiple middle section mining (two)

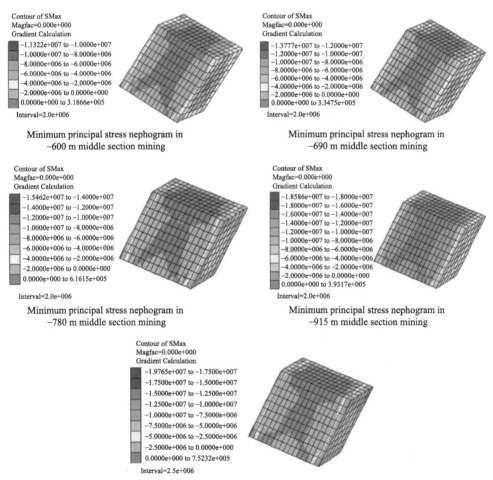

Figure 6-17 Distribution law of minimum principal stress in deep multiple middle section mining

Maximum principal stress and minimum principal stress extremum
of simulated mining in deep middle section   Table 6-8

Middle section depth(m)	Maximum principal stress(MPa)	Minimum principal stress(MPa)
600	37.331	11.322
690	39.844	13.777
780	43.393	15.462
915	49.897	18.586
1050	52.024	19.586

Figure 6-18 is the nephogram of plastic zone after simulated mining in deeper horizontal multiple middle section. The simulation results show that the distribution laws of plastic zone, shear failure and tension failure are similar in each middle section. However, with the increase of mining depth, the overall dam-

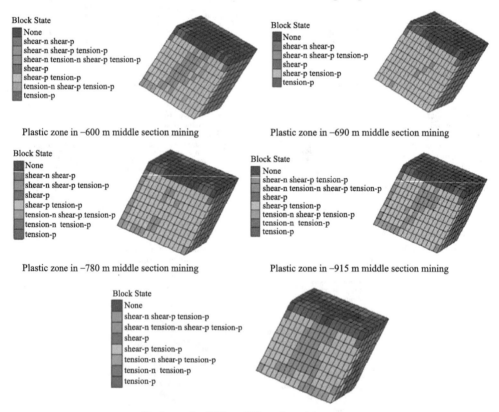

Figure 6-18 Distribution law of plastic zone in deep multiple middle section mining

age range tends to increase. In the actual deep mining, attention shall be paid to the damage of rock mass.

6.5 Stability analysis of surrounding rocks in deep fault zone

F3 fault, which runs through the mining area directly under Sanshandao Gold Mine, is a difficult fault encountered in the process of mine roadway development and construction. The average width of fractured zone is about 10-30 m, with abundant groundwater reserves. The long-term immersion of groundwater in the fractured zone makes the physical and mechanical properties of rocks and fault gouges in the zone worse, which further reduces the mechanical properties of the already broken rocks and threatens the safe construction of mine development projects. Due to the large scale of F3 fault and its extension in the deep area of the mine, the development project of the transportation roadway in the deep middle section of Gold Mine will inevitably be affected by F3 fault. Therefore, how to make a scientific design, construction and support scheme, so that the roadway development process can safely pass through F3 fault zone and serve the mine production stably for a long time, is a long-term focus of Sanshandao Gold Mine.

6.5.1 Support design and construction scheme of −690 m north roadway F3 fault zone

The excavation engineering in −690 m north roadway through F3 fault zone is planned to adopt the scheme of pipe shed pre-support. According to previous construction experience, hazards such as water burst, roadway roof and collapse of two sides are easy to occur in the construction process, which makes the construction difficult. The construction can be carried out under the conditions of safe protective measures and advanced construction technology. In addition, all construction links shall be closely linked to ensure that all operations including roadway excavation, U-shaped steel support and shotcrete support are completed in the shortest time. The concrete process of construction technology is divided into the following stages:

(1) Before excavation, the U-shaped steel support on the surface is checked, and the quality of the U-shaped steel and whether the parts are neat is

checked.

(2) Excavation engineering is carried out, and support is used to assemble chamber; The U-shaped steel shall be loaded at the same time of excavation, and the U-shaped steel shall be transported to the underground.

(3) After the roadway construction of the assembled chamber, the supporting operation shall be immediately carried out, firstly, the support is connected, then they are pushed horizontally to the support area with a scraper, the number of steel supports is determined on site every time, then the supports are fixed immediately, the steel mesh and straw curtain are quickly installed, and the two sides and the vault are quickly filled and compacted with timber and waste rock to complete the supporting operation.

(4) After the steel support is finished, the concrete arch support shall be carried out quickly. After the concrete arch support is finished, the vault gap shall be filled with cement mortar by shotcrete machine, and the two sides will be supported by shotcrete. Concrete spraying will be carried out twice in areas with arches. The first initial spraying of 3-5 cm is used to close the surrounding rock as early as possible, fill the rock surface, and then lay the anchor rods, and then re-spray after erecting the arch support to meet the design thickness requirements. For the rock surface with water leakage, water diversion treatment shall be made before concrete spraying, and then concrete spraying shall be carried out, and the water shall be approached from the place with water to the place without water.

(5) After the excavation and supporting construction of the support assembly chamber is completed, the anchor rod shall be pre-supported according to the design and construction. Bolt with circumferential spacing of 0.2 m, length of 5 m and diameter of 32 mm is used for pre-support.

(6) After the pre-supporting bolt is finished, the broken zone is excavated immediately. Firstly, the small section passes through the broken zone, then on-site technical analysis is carried out, and then 2-5 m is advanced according to the designed section of roadway. Finally, roof caving and supporting operation is carried out in the broken zone section. The section of roadway is $5.2 \text{ m} \times 4.5 \text{ m}$.

(7) Circulation construction from step (1) to step (6) is carried out until it passes through F3 fault fracture zone. The following key points shall be paid attention to in the construction: less disturbance, that is, reducing the frequency and intensity of disturbance of surrounding rock, trying to blast the whole

section as much as possible, and controlling the dosage; Fast evacuation, that is, the evacuation speed must be fast, and the evacuation g task of the fault zone shall be completed within 24 hours; Quick support, that is, U-shaped steel support shall be carried out immediately after the excavation, and this work shall be completed within 8 hours; Tight sealing, that is, spraying concrete and grouting in time to shorten the time of surrounding rock exposed to air.

The supporting design scheme of the above construction process is shown in Figure 6-19.

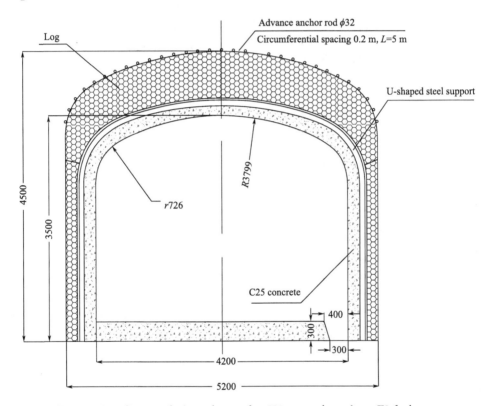

Figure 6-19 Support design scheme of −690 m north roadway F3 fault zone

6.5.2 Numerical model for the roadway evacuation

The numerical calculation model of the zone near F3 fault in the north roadway in −690 m middle section is established by using ANSYS and FLAC3D software. In the model, X axis is taken as the horizontal direction perpendicular to the roadway strike, Y axis is taken as the parallel direction of roadway strike, and Z axis is taken as the vertical direction. The length of the model in the direction of three coordinate axes is 50 m, and it is divided into two blocks: foot wall surrounding rock group and fault group (with the thickness of 10 m), which are divided into 163788 units and 28441 nodes, as shown in Figure 6-20.

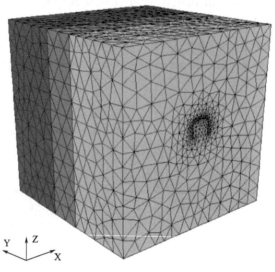

Figure 6-20　Numerical calculation model of −690 m north roadway excavation engineering through F3 fault zone

As there is abundant groundwater stored in F3 fault zone, the surrounding rock in the surrounding area is under the action of highly mineralized and weakly acidic groundwater for a long time. Considering the hydrochemical damage effect of granite, as groundwater also has certain erosion effect on steel structure and concrete, the damage effect on supporting materials shall also be considered in the calculation process. Roadway excavation is simulated by distributed excavation, which is carried out every 5 m for 10 times in total.

6.5.3　Displacement field analysis

The Z-direction displacement field of surrounding rock around excavated roadway obtained by simulation calculation is shown in Figure 6-21 to Figure 6-24. According to the analysis of roadway cross-section displacement diagrams at different positions in the fault zone, the deformation of the roof and two sides of the roadway has been effectively controlled due to the advanced anchor cable support, log cushion, U-shaped steel frame and shotcrete support in the process of roadway excavation: the maximum deformation in the fault zone is within 2 cm, and that in the vicinity of the fault is 6 mm. In the same cross-section position, the maximum displacement occurs at the roadway floor, and the displacement direction is upward. The maximum displacement of floor in non-fault zone is 2 cm, and the maximum displacement of floor in F3 fault zone reaches 8 cm. This is because the stress can only be released on the floor with weak support in the process of roadway development, which leads to a large displacement of floor. Therefore, attention shall be paid to the displacement at the floor in the process of roadway development. In areas with obvious deformation and easy damage, appropriate support measures shall be taken, and closed rings shall be added when conditions permit to enhance the integrity of roadway support system.

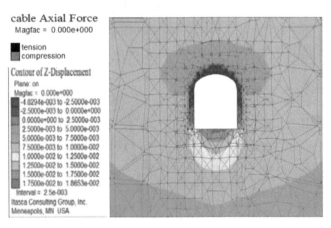

Figure 6-21　Z-direction displacement at 5 m from F3 fault

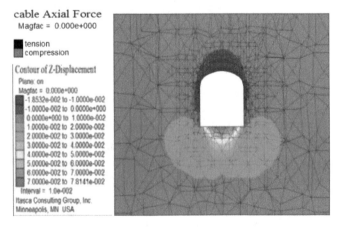

Figure 6-22　Z-direction displacement of the first excavation of F3 fault

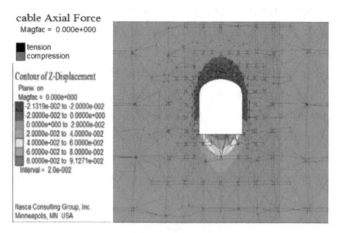

Figure 6-23　Z-direction displacement of the second excavation of F3 fault

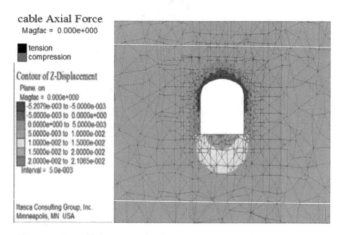

Figure 6-24　Z-direction displacement at 5 m through F3 fault

6.5.4　Plastic failure zone analysis

The Z-direction displacement field of the surrounding rock around excavated roadway obtained by simulation calculation is shown in Figure 6-25 to Figure 6-28. Through the distribution map of plastic failure zone, it can be found that the support system with U-shaped steel frame and advanced anchor cable as the main body can provide reliable support when the rock mass around the roadway is damaged, and control the plastic failure zone of the roof and two sides of the roadway within a reasonable range. Compared with the Z-direction displacement nephogram, it can be found that the plastic failure zone is mainly concentrated in the weak support floor.

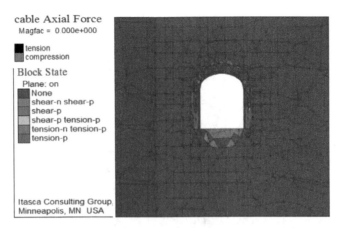

Figure 6-25 Plastic failure zone at 5 m away from F3 fault

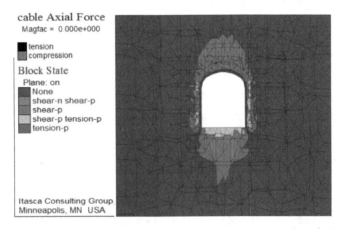

Figure 6-26 Plastic failure zone in the first excavation of F3 fault

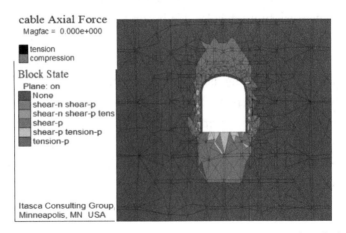

Figure 6-27 Plastic failure zone in the second excavation of F3 fault

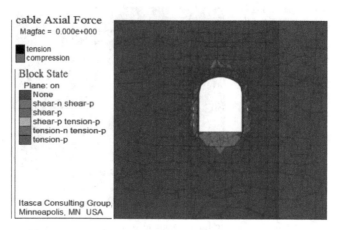

Figure 6-28 Plastic failure zone at 5 m through F3 fault

By analyzing the numerical simulation results of roadway development process, it can be found that the advanced bolt is embedded in the surrounding rock, and the surrounding rock is connected to form an approximate shell structure. In the process of roadway excavation, the bolt shall bear most of load and play an important role in supporting the surrounding rock. After the U-shaped steel frame is fixed, concrete is sprayed immediately to close the exposed surface of surrounding rock, and the solidified concrete can directly provide the arch support resistance to the surrounding rock; In addition, concrete grout infiltrates into the open cracks or joints of surrounding rock, which improves the strength of surrounding rock, and bonds the U-shaped steel frame with the surrounding rock to enhance its supporting effect. It can be seen that the supporting system composed of advanced support bolt, U-shaped steel frame and sprayed concrete has a good supporting effect on the surrounding rock near F3 fault zone in -690 m middle section north roadway, and can meet the requirements of safe construction of development engineering and service of the roadway for mine production. However, the deep of mining area is under high geostress and the hydrogeological conditions are more complex. In the field construction of mine and roadway engineering, attention shall be paid to the possible problems of roadway floor heave, and the monitoring of surrounding rock deformation during construction shall be strengthened.

7

Stability Analysis and Ground Pressure Monitoring

7.1 Three-dimensional fracture network model

7.1.1 Establishment of the model

Through the measurement of local outcrop of rock mass or outcrop area caused by excavation by three-dimensional non-contact scanning photography technology and later statistics, the model with the same probability distribution characteristics as the real distribution of joints and fissures is obtained, and then the size of a certain three-dimensional space area can be determined.

Taking the surrounding rock of tunnels in Sanshandao Gold Mine as the analysis object, a three-dimensional joints and fissures network model with a spatial dimension of 20 m×20 m×20 m is established, in which the spatial coordinates of 3DEC system take due east and due north vertical directions as X-axis, Y-axis and Z-axis respectively. The three-dimensional fissures network model is generated according to the fissures parameters investigated and statistics in Chapter 3. The DFN model and profile are shown in Figure 7-1.

7.1.2 Reconstruction of fractured rock mass

The reconstruction of fractured rock mass is to embed the DFN model into the complete rock block model for cutting, so as to obtain a discrete block model with the same or close fissures probability distribution characteristics as the actual rock mass. By constructing a 5 m×5 m×5 m rock block and then importing into the DFN joints and fissures model for cutting, an example can be shown. The method of fractured rock mass reconstruction is shown in Figure 7-2.

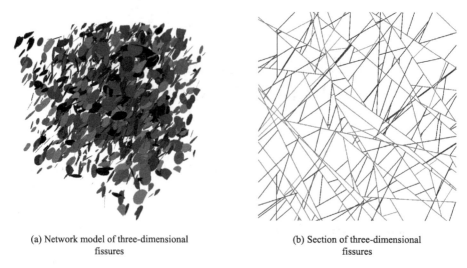

(a) Network model of three-dimensional fissures

(b) Section of three-dimensional fissures

Figure 7-1　Network model and section of three-dimensional fissures

Figure 7-2　Reconstruction process of fractured rock mass

7.2　Mechanical parameters of representative elementary volume

The selection of location points on site is extremely restrictive to the determination and acquisition of rock mass mechanical parameters, and it is difficult to measure and obtain rock mass mechanical parameters at the test site. In addition, because the occurrence condition of rock mass has a variety of variability, it is difficult to accurately summarize the characteristics of rock mass. It has been proved by practice that all kinds of empirical methods are not simple and easy to use, and how to obtain relatively accurate mechanical parameters of rock mass according to indoor rock mechanics tests and distribution laws of joints and fissures has been an urgent and difficult engineering problem to solve.

There are joints, bedding and other defects ranging in size from centimeters to meters occurring in the engineering rock mass. Such defects are distributed in a wide range and extremely random, and even there are some large-scale defects such as faults and dikes in the rock mass, whose size can extend tens to hundreds of meters. In a small scope of engineering, the engineering rock mass is a discontinuous medium material. However, under certain engineering conditions, the rock mass can still be regarded as a continuous medium material, so the continuum mechanical analysis of rock mass can be carried out from the macro level.

Rock mass can be regarded as a material composed of rock mass micro-elements, and the size of micro-elements constituting rock mass can be different. When the size of rock mass micro-elements increases from small to a certain critical size, rock mass will show certain mechanical properties of continuous medium. At this time, the engineering fractured rock mass composed of micro-elements of this size can be regarded as continuous medium material, and the micro-elements of rock mass is the representative elementary volume of rock mass, and the micro element size is the size of minimum representative elementary volume. When the size of the minimum representative elementary volume is determined, the equivalent mechanical parameters of rock mass can be obtained, and the rock mass can be regarded as a continuous medium for analysis. As a continuous medium material, it is considered that the mechanical properties of the rock mass can follow the elastic-plastic theoretical properties of continuous medium mechanical materials. To a certain extent, the representative elementary volume determines whether the rock mass can be regarded as a continuous medium material approximately, so the research on the mechanical properties of rock mass is of great value. When the size of the rock mass model is close to or even larger than the size of the representative elementary volume, the mechanical parameters of the rock mass model can be regarded as the mechanical parameters of the equivalent complete rock material.

Based on the distribution characteristics of joints and fissures probability measured on site, and according to the distribution parameters of joints and fissures, a three-dimensional fracture network model (DFN) is constructed. On this basis, the DFN model is embedded in the block and cut to obtain the fractured rock mass model, combined with the three-dimensional block discrete element model 3DEC to establish the DFN-DEM coupling equivalent rock mass model. Besides, the fish function is used to control the uniaxial and triaxial compression numerical tests, and quantitative studies on the anisotropy and strength of the equivalent rock mass are carried out, so as to obtain the minimum representative elementary volume of rock mass, obtaining the equivalent rock mass mechanical parameters such as tensile strength, cohesion, internal friction angle, elastic modulus and Poisson's ratio.

7.2.1 Determining the representative elementary volume

In order to analyze the size of the rock mass representative elementary volume, the sizes of 1 m×1 m×1 m, 2 m×2 m×2 m, 3 m×3 m×3 m, 4 m×4 m×4 m, 5 m×5 m×5 m and 6 m×6 m×6 m concentric cubes were established respectively model of fractured rock mass. Three sets of tests were performed for each group of models, and the uniaxial compression numerical tests were performed on the X-axis, Y-axis, and Z-axis directions to obtain the X, Y and Z direction elastic modulus and peak strength of models of different sizes. The models are shown in Figure 7-3. See Table 7-1 and Table 7-2 for the mechanical parameters of rock blocks and structural planes.

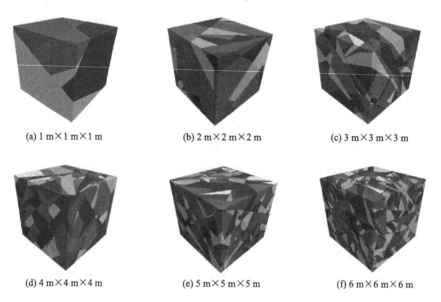

(a) 1 m×1 m×1 m (b) 2 m×2 m×2 m (c) 3 m×3 m×3 m

(d) 4 m×4 m×4 m (e) 5 m×5 m×5 m (f) 6 m×6 m×6 m

Figure 7-3 Equivalent rock mass models of different sizes

7 Stability Analysis and Ground Pressure Monitoring

Mechanical parameters of rock blocks   Table7-1

Elastic modulus (GPa)	Poisson's ratio	Cohesion (MPa)	Internal friction angle (°)	Tensile strength (MPa)
43.98	0.2	17.19	49.96	6.93

Mechanical parameters of structural plane   Table 7-2

Normal stiffness (GPa)	Shear stiffness (GPa)	Cohesion (MPa)	Internal friction angle (°)	Tensile strength (MPa)
150	110	15	28	4

In uniaxial compression test, all nodal forces at the top of the specimen are extracted by fish language, and the axial stress is obtained by dividing the superposition of nodal forces at the top by the area of the top. Record the top displacement, and divide the top displacement by the size of the specimen to get the strain, thus obtaining the stress-strain curve and obtaining the mechanical properties of rock mass according to the stress-strain curve.

According to the numerical test results, the stress-strain curves of elastic modulus of X-axis, Y-axis and Z-axis with different size models under uniaxial compression are shown in Figure7-4.

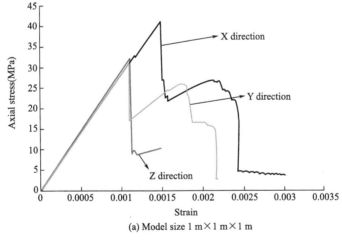

(a) Model size 1 m×1 m×1 m

Figure 7-4 Uniaxial compression stress-strain curves of equivalent rock masses of different sizes (one)

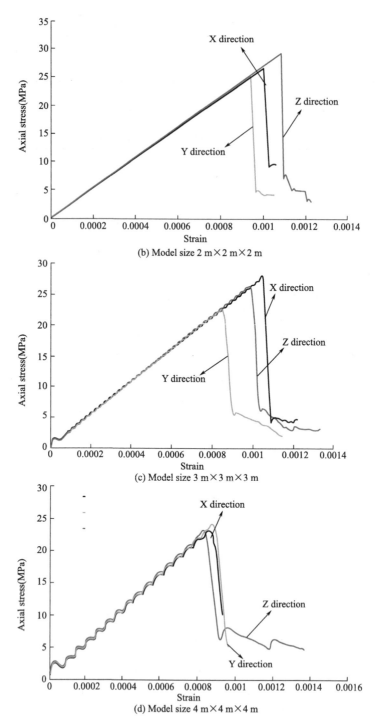

Figure 7-4 Uniaxial compression stress-strain curves of equivalent rock masses of different sizes (two)

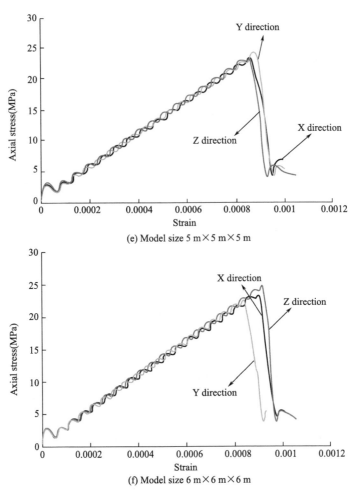

Figure 7-4 Uniaxial compression stress-strain curves of
equivalent rock masses of different sizes (three)

As shown in Figure 7-4, before the uniaxial compression peak strength of the rock mass, the relationship between stress and strain is basically linear. When the stress increases to the peak strength, the stress drops sharply with the strain. In terms of numerical test results, the fractured rock mass is a brittle material under uniaxial compression.

Uniaxial compression numerical tests of the same size are carried out along X, Y and Z directions, and the total stress-strain curves, peak strength and elastic modulus in each direction are obtained. The mechanical properties in different directions are different, which is due to the anisotropy caused by the existence of joints and fissures in the rock mass model. When the model size is small, the anisotropy characteristics of the rock mass are shown in Figure 7-4 (a) - (c). The stress-strain curves in the three directions are quite different, showing extremely obvious anisotropy. As shown in Figure 7-4 (d) - (f), when the model size is larger and larger, the difference of stress-strain curves in three directions becomes smaller, and the anisotropic characteristics become weaker.

In order to obtain anisotropic characteristics of different sizes in elastic stage, according to the stress-strain curves of each group of tests, the variation trend of elastic modulus with model size in X, Y and Z directions is extracted. The curves of elastic modulus in different model sizes are shown in Figure 7-5, and the uniaxial peak strength in different directions is shown in Figure 7-6.

According to the change curve of peak strength and elastic modulus of jointed rock mass with model size, it shows the law that the peak strength and elastic modulus decrease with the increase of model size, and finally tend to be stable, but there are local fluctuations, which is caused by the randomness of joint distribution. When the number of joints is small, the joints' position has an important influence on the model, and the edge-fissures have a greater influence on the overall strength of rock mass than the internal-fissures; It can be seen from the peak strength and elastic modulus in X, Y and Z directions of the jointed rock mass model with the same size that the jointed rock mass has anisotropic mechanical properties. However, with the increase of the size, the seed delivery gap gradually decreases, and the anisotropy gradually weakens, that is, with the increase of the size of the model, the number of cracks increases. The rock blocks are cut by joints from all directions, and the rock blocks can be approximately regarded as homogeneous and isotropic; With the increase of equivalent rock mass model size, the peak strength and elastic modulus in X, Y and Z directions gradually decrease and eventually become stable, indicating the existence of representative elementary volume in jointed rock mass. Based on the above analysis results, the representative elementary volume is taken as 6 m×6 m×6 m, and the average value showed that the elastic modulus at this scale is 25.3 GPa and the uniaxial compressive strength is 24 MPa.

7.2.2 Calibration of mechanical parameters of equivalent rock mass

According to the change curve of peak strength and elastic modulus of jointed rock mass with model size, it shows the law that the peak strength and elastic modulus decrease with the increase of model size, and finally tend to be stable.

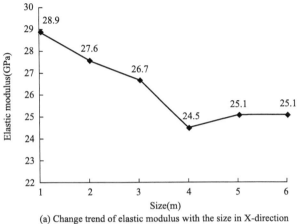
(a) Change trend of elastic modulus with the size in X-direction

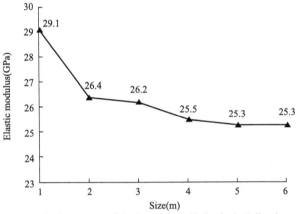
(b) Change trend of elastic modulus with the size in Y-direction

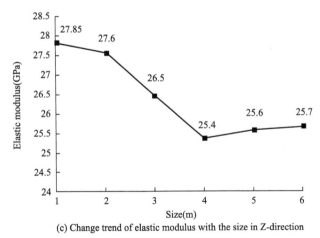
(c) Change trend of elastic modulus with the size in Z-direction

Figure 7-5　Change trend of elastic modulus with the size in different directions

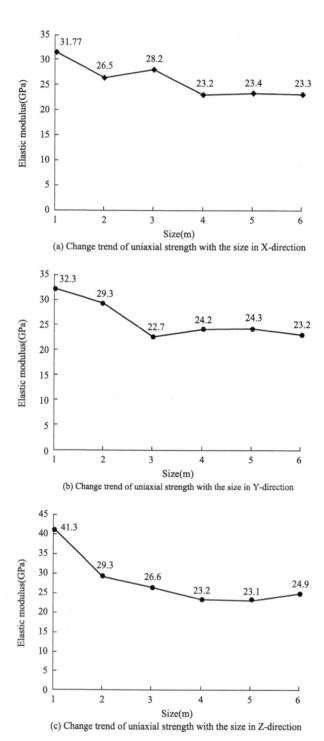

Figure 7-6　Change trend of uniaxial compression strength with the size in different directions

When the content of joints and fissures is small, the location of fissures occurrence has a direct impact on the calculation results of the model, resulting in a certain fluctuation of the curve. Comparing the values of peak strength and elastic modulus in three directions of the single size model, it can be seen that the anisotropy of fractured rock mass gradually changes into isotropy with the increase of the size of the rock mass. This is due to the fact that with the increasing size of the model, there are more and more joints and fissures in the rock mass, and the rock block is cut by random structural planes in all directions. According to the above analysis, the minimum representative elementary volume size can be taken as 6 m×6 m×6 m, in which the average elastic modulus is 25.3 GPa and the peak strength is 24 MPa. On the basis of determining the size of representative elementary volume REV, the mechanical parameters such as cohesion, internal friction angle and tensile strength of equivalent rock mass can be calculated through corresponding triaxial compression numerical test.

In this triaxial compression numerical test, the model of 6 m×6 m×6 m is used to carry out the triaxial compression test under equal confining pressure, and the loading confining pressure is 17 MPa, and the obtained stress-strain curve is shown in Figure 7-7. According to the obtained curve, the compressive strength of rock mass is 49.8 MPa under the action of equal confining pressure of 17 MPa, indicating that the compressive strength of rock mass increases under the condition of confining pressure. According to the curve after peak, rock mass shows strong ductility under three-dimensional stress condition, and fractured rock mass changes from brittle material to ductile material.

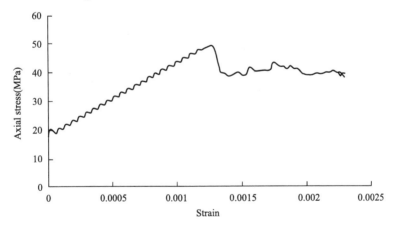

Figure 7-7　Triaxial compression stress-strain curve

According to the Mohr-Coulomb criterion, the equivalent rock mass mechanical parameters are calculated as shown in Table 7-3.

Equivalent rock mass mechanical parameters Table 7-3

Elastic modulus (GPa)	Poisson's ratio	Cohesion (MPa)	Internal friction angle(°)	Tensile strength (MPa)
25.3	0.3	6.93	30	1.53

7.3 Stability analysis of the tunnel in fracture rock mass

7.3.1 Establishment of three-dimensional model of horizontal tunnel

According to the CAD plane development drawing of Xishan Branch of Sanshandao Gold Mine, the geometric shape and direction of −870 m horizontal tunnel are analyzed. The CAD plan layout is shown in Figure 7-8.

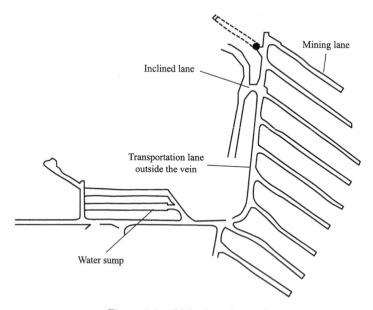

Figure 7-8 CAD plan of tunnel

In order to facilitate the analysis of the overall stability of the −870 m horizontal tunnel, the overall modeling method is proposed to establish the three-dimensional geometric model of the tunnel through the three-dimensional modeling software Rhinoceros. The modeling commands that are used mainly include: plane stretching, sweeping, lofting, uniting and moving, etc. The main steps of modeling: first, determine the coordinates of the two ends of the tunnel, determine the specific location of the tunnel, and establish the tunnel wire frame. Second, carry out the operation of lofting, stretching, sweeping the wire frame to obtain the entity model of tunnel. And finally, the tunnel model is combined with the calculation method of union in the Boolean operation to form the overall model of the horizontal tunnel. The tunnel model is shown in Figure 7-9.

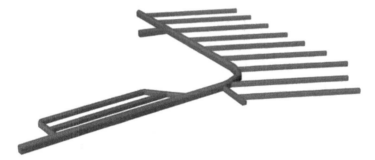

Figure 7-9 Rhinoceros three-dimensional geometric model of tunnel

The main steps of establishing numerical model: based on the Rhinoceros three-dimensional model, the tunnel geometric model is meshed through the plug-in Griddle, and the grid nodes are exported in the file format of f3 grid, and the file is imported into FLAC3D to form a -870 m horizontal tunnel numerical model. The numerical model is shown in Figure 7-10, and the model parameters are shown in Table 7-4.

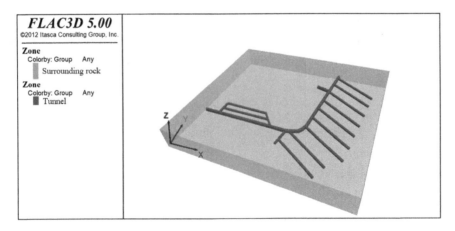

Figure 7-10 FLAC3D numerical model

Table of tunnel numerical model parameters	Table 7-4
Length in X-direction(m)	278
Length in Y-direction(m)	250
Length in Z-direction(m)	35
Number of model nodes	472079
Number of model elements	2676174

7.3.2 Mechanical model of −870 m horizontal tunnel

1) Constitutive model and failure criterion of rock mass

The mechanical analysis process of deep rock mass often adopts isotropic elastoplastic constitutive structure, and the main failure criterion is the Mohr-Coulomb failure criterion. The Mohr-Coulomb failure criterion is expressed as follows: when the stress distribution of rock and soil materials reaches the strength limit, the tangential stress on a certain plane of rock and soil materials will exceed the shear strength of that plane, resulting in plastic flow failure. According to the laws of elastoplastic mechanics of rock and soil, a series of mechanical tests are carried out on rock and soil materials: uniaxial tensile strength test, uniaxial compressive strength test, and triaxial compression strength test. According to each test, the Molar circle under different loading conditions can be obtained, thus obtaining the envelope of the Mohr circle of the material. It is generally believed that the Mohr circle of rock and soil can be connected by a straight line when the stress change value is small, and the mechanical properties obtained by this method are quite different from the actual ones.

The failure criterion combining the Coulomb straight line failure criterion and the Mohr failure theory is called the Mohr-Coulomb criterion. The criterion is shown in equation (7-1), and the element yield stress expression is shown in equation (7-2):

$$\tau = c + \tan\varphi \qquad (7\text{-}1)$$

$$f = \sigma_3 - N_o \sigma_1 + 2c[(1+\sin\varphi)/(1-\sin\varphi)]^{\frac{1}{2}} \qquad (7\text{-}2)$$

where c—cohesion;
τ—limit shear stress;
φ—internal friction angle;
σ—plastic surface normal stress;
σ_1—maximum principal stress;

σ_3—minimum principal stress.

By calculating the f value, FLAC3D can judge whether the element has plastic failure, and its judgment equation is:

$$f = \begin{cases} >0 & \text{Element yields} \\ <0 & \text{Element doesn't yield} \end{cases} \quad (7\text{-}3)$$

The failure criterion adopted is Mohr-Coulomb failure criterion, which is one of the built-in failure criteria of FLAC3D numerical software, and can accurately reflect the mechanical strength properties of rock mass.

2) Selection of mechanical parameters and model boundary conditions of rock mass materials

According to the distribution characteristics of joints and-fissures and complete rock combination analysis, the size of representative elementary volume and mechanical parameters are determined, and the mechanical parameters of the representative elementary volume are shown in Table 7-5.

Mechanical parameters of equivalent rock mass   Table 7-5

Elastic modulus (GPa)	Poisson's ratio	Cohesion (MPa)	Internal friction angle(°)	Tensile strength (MPa)
25.3	0.3	6.93	30	1.53

The boundary conditions of the numerical calculation model are determined according to the in-situ stress measurement: based on the measured in-situ stress distribution, a straight line is fitted to each horizontal in-situ stress value, and the gradient of the in-situ stress change is the slope of the straight line. According to the initial in-situ stress measurement level from shallow to deep, the −870 m horizontal in-situ stress distribution characteristics can be calculated and analyzed according to the linear equation, and the horizontal maximum and minimum horizontal principal stress and vertical stress can be obtained. The boundary conditions of the numerical model can be obtained by conversion according to factors such as azimuth and inclination.

The in-situ stress can be calculated: maximum horizontal principal stress $\sigma_{yy}=40.11$ MPa, minimum horizontal principal stress $\sigma_{xx}=21.52$ MPa, vertical stress $\sigma_{zz}=25$ MPa.

7.3.3 Stability analysis of surrounding rock in −870 m horizontal tunnel

1) Determination of maximum unbalance force monitoring and model size

The simulation sets the unexcavated model as an isotropic elastic body, and calculates the in-situ stress value as the initial stress of the model. After the stress is balanced, the displacement and velocity of the model are cleared, and then the tunnel is excavated, and the program is run again to balance. The running time of the process is 2070 steps. After the model reaches equilibrium, the constitutive definition of the model is redefined, the complete elastic body is transformed into an elastoplastic body, the failure criterion is the Mohr-Coulomb criterion, and then the equilibrium is performed again. The running time of the process is 550 steps. Figure 7-11 shows the monitoring curve of the maximum unbalanced force over time.

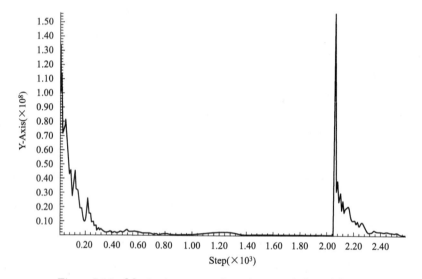

Figure 7-11　Monitoring curve of maximum unbalanced force

Since the boundary condition of the model is set to in-situ stress, the stress from the inside of the model to the geometric boundary is required to be gentle and excessive, and when it reaches the geometric boundary of the model, the stress should be gentle and excessive to the in-situ stress value, indicating that the model size is reasonable. On the basis of reasonable model size, reducing the model size as much as possible is beneficial to the running speed of the program. In order to analyze the rationality of the model size, two monitoring lines are arranged in the model, and the distribution of the maximum, minimum and intermediate principal stresses on each monitoring line can be obtained through fish language programming. The monitoring line layout is shown in Figure 7-12.

The distribution curves of the maximum, minimum, and intermediate prin-

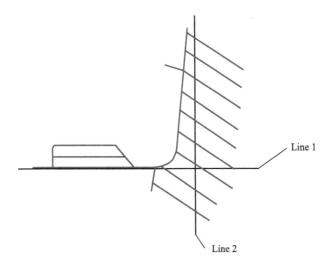

Figure 7-12 Layout position of monitoring lines

cipal stresses monitored by monitoring Line 1 and Line 2 are shown in Figure 7-13 and Figure 7-14. In FLAC3D system, the pressure is negative, and the maximum principal stress on each measuring line is gently over −40.11 MPa, the intermediate principal stress is gently over −25 MPa, and the minimum principal stress is gently over −21.5 MPa. According to the distribution of maximum, minimum and intermediate principal stress on the monitoring curve, it can be judged that it is more reasonable that the size in the X-direction is 278 m, the size in the Y-direction is 250 m, and the size in the Z-direction is 35 m.

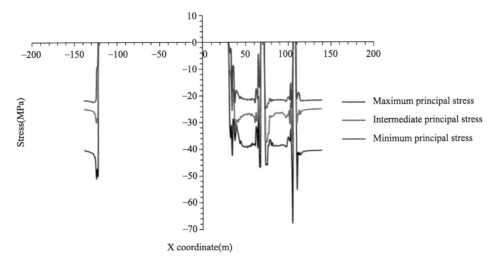

Figure 7-13 Principal stress distribution curve on monitoring Line 1

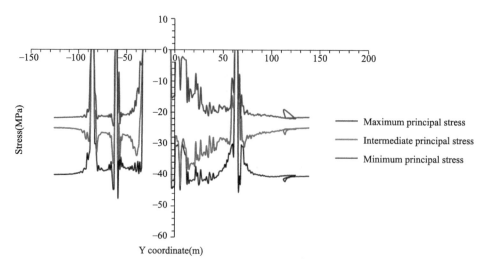

Figure 7-14   Principal stress distribution curve on monitoring Line 2

Figure 7-15   Roof breakages in −870 m horizontal mining-connected tunnel roof breakage

2) Analysis of the instability area of the surrounding rock of tunnels

After the −870 m horizontal tunnel is opened, it is only partially supported by pipe joint bolts, and the overall tunnel is not permanently and effectively supported. Local fragmentation and roof falling safety incidents have occurred in the tunnel, as shown in Figure 7-15. In particular, roof falling accidents have occurred at the entrance of the slope-connected tunnel and mining-connected tunnel. The destruction and instability of the surrounding rock of the tunnel seriously threatens the safety of workers and vehicle transportation. It is urgent to analyze the dangerous location of the surrounding rock of the tunnels qualitatively and quantitatively and to support it.

After the tunnel excavation is balanced, the model element is changed from isotropic complete elasticity to elastic-plastic element, and the program is run to solve it iteratively. After the maximum unbalanced force tends to be stable, the plastic element is extracted and displayed separately by programming fish language, and the plastic element is counted to visualize the plastic failure area of the model, so that the vulnerable position of surrounding rock of tunnels can be controlled macroscopically, so as to support the tunnel in the later stage. The distribution of plastic elements of surrounding rock of tunnels is shown in Figure 7-16. The statistics of plastic failure elements of tunnel surrounding rock show that there are 2319 plastic elements in total.

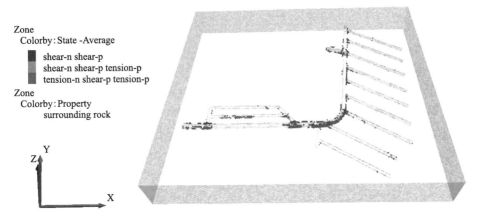

Figure 7-16 Distribution of plastic elements in the surrounding rock of tunnels

According to the judgment of the distribution position of theplastic elements, the two sides of the slope-connected tunnel and the roof have different degrees of damage, and the main form of failure is shear failure. Since the slope-connected tunnel is the throat position that connects the slope and the transportation tunnel outside the vein, the staff and vehicles must pass through the slope-connected tunnel when entering the transportation tunnel, so this location is more important. Therefore, the slope-connected tunnel needs to be supported and reinforced to ensure the safety of workers and the smooth flow of transportation.

The plastic element is relatively concentrated at the connection position of the transportation tunnel outside the vein and mining-connected tunnel, while the surrounding rock of the mining-connected tunnel is rarely damaged. But the entrance of the mining-connected tunnel is not only the entrance of the staff to the working face, but also the entrance of the ore handling vehicles. The instability of the surrounding rock at this location poses a certain threat to the normal safety of production, so it is necessary to carry out local reinforcement treatment at the entrance of mining-connected tunnel.

In addition to the concentrated damage at the entrance of the slope-connected tunnel and the mining-connected tunnel, the most serious damage area is at the corner of the transportation tunnel outside the vein, and the east-west section is longer, which is about 60 m long. This position is the transition section of transportation tunnel from north-south to east-west, and it also intersects with some mining-connected tunnel. The instability of surrounding rock at this position not only seriously affects the safety of workers, but also seriously affects the transportation.

3) Stress analysis of surrounding rock of tunnels

In order to analyze the stress distribution of surrounding rock of tunnels, the model stress distribution is analyzed from two aspects of minimum principal stress and maximum shear stress. According to the distribution of the minimum principal stress, the distribution of the tensile stress can be judged. When the minimum principal stress at a certain position is positive, the tensile failure may occur at that position. According to the distribution of the maximum shear stress, the most vulnerable position of shear failure can be determined, and the model can be sliced. Through slicing, the model can be viewed from the top and bottom, which can comprehensively reflect the stress distribution of the top and bottom.

The distribution of the minimum principal stress of the roof is shown in Figure 7-17. The maximum value of the maximum horizontal principal stress of the roof position is −0.6 MPa, and the overall minimum principal stress of the roof position is shown as compressive stress. There is basically no tensile stress in the

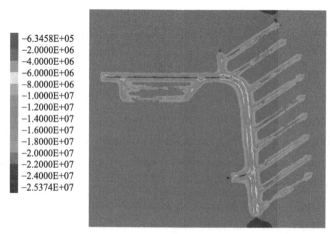

Figure 7-17 Bottom view of the minimum principal stress nephogram slice

roof position, indicating that the probability of tensile failure of the roof is small. However, when the dangerous block is suspended from the roof, attention should also be paid to the possibility of tensile failure of the block under the action of dead weight stress, resulting in the block falling, which poses a threat to the staff or transport vehicles.

The top view of the minimum principal stress nephogram slice is shown in Figure 7-18. The positions with the minimum principal stress value > 0 are mainly concentrated in the intersection position of the mining-connected tunnel and the transportation tunnel outside the vein, and in the position of the tunnel floor, and the minimum principal stress is the most concentrated at the intersection of the slope-connected tunnel and the transportation tunnel outside the vein. It should be noted that these locations are more prone to rib spalling.

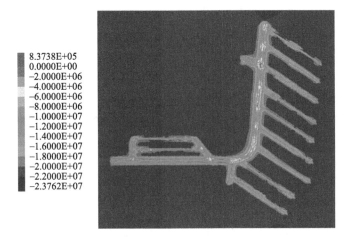

Figure 7-18 Top view of the minimum principal stress nephogram slice

In order to make the area where the minimum principal stress is tensile stress appear more intuitively, the elements with the minimum principal stress >0 are extracted by fish language programming, and these elements are grouped and displayed separately, thus realizing three-dimensional visualization of tensile stress distribution position. The distribution of tension elements is shown in Figure 7-19. There are 801 tension elements in total, and the parts with tensile stress are mainly distributed at the intersection of slope-connected tunnel and transportation tunnel, the position of sump and the corner of transportation tunnel outside the vein. It should be noted that the tensile strength of the contact surface of general fissures is low, and these locations are prone to secondary ten-

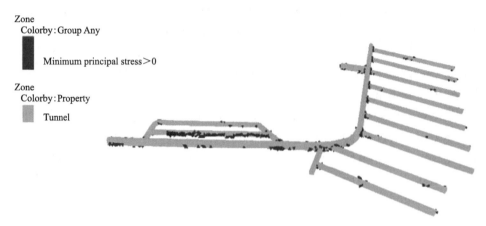

Figure 7-19 Distribution of tensile stress element

sile failure of the primary fissures due to the existence of tensile stress.

The difference between the maximum principal stress and the minimum principal stress of each element is extracted by programming with fish language, and then the maximum shear stress of each element is calculated, and the maximum shear stress is assigned to each element by means of additional variables, so as to obtain the nephogram of shear stress. The calculation method is as follow:

$$\sigma_{shear} = abs\left[\frac{(\sigma_1 - \sigma_3)}{2}\right] \quad (7-4)$$

where σ_{shear}—maximum shear stress;
abs [ ] —absolute value.

The bottom and top views of the maximum shear stress nephogram slice are shown in Figure 7-20 and Figure 7-21. According to the nephogram distribution of the maximum shear stress of the roof, the maximum shear stress is mainly concentrated in the slope-connected tunnel and the corner of the tunnel, and the maximum shear stress at the corner reaches 30 MPa; According to the distribution of the maximum shear stress nephogram in the top view of the model, the maximum shear stress is mainly in the slope-connected tunnel and the east-west transportation tunnel. Because the rock mass is the one with joints and fissures, the mechanical properties of tunnel surrounding rock are weaker than that of intact rock, and their mechanical properties largely depend on the mechanical properties of structural planes. Under the action of shear stress, the structural plane is easy to slip, which leads to the failure of surrounding rock. According to the above analysis, the locations where block shear slip easily occurs in the roof are

■ 7 Stability Analysis and Ground Pressure Monitoring ■

mainly concentrated at the corner between the slope-connected tunnel and the tunnel, so it should be prevented that the falling of the roof block poses a threat to the staff and transportation vehicles; The maximum shear stress of the east-west transportation tunnel is mainly concentrated on two sides of the tunnel, so it should be noted that the blocks of two sides of the tunnel fall off, which hinders the transportation vehicles.

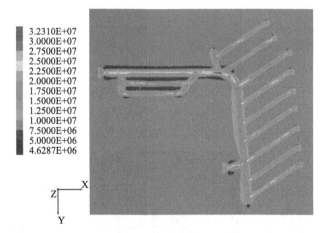

Figure 7-20　Bottom view of the maximum principal stress nephogram slice

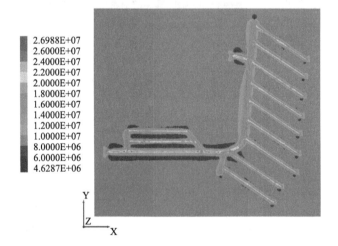

Figure 7-21　Top view of the maximum principal stress nephogram slice

4) Deformation analysis of surrounding rock of tunnels

In order to analyze the tunnel deformation, the horizontal displacement of the tunnel model and the subsidence displacement of roof are analyzed respective-

211

ly. The displacement in the horizontal direction of the tunnel model is mainly reflected in the displacement of two sides of the tunnel. The displacement is mainly represented by the horizontal displacement cloud diagram of two sides of the tunnel. The tunnel model is sliced to observe its horizontal displacement changes. In order to comprehensively reflect the horizontal displacement of tunnel in all directions, each node is assigned with the total displacement of tunnel as an additional variable by programming with fish language, and the calculation method is as follow:

$$dis_h = \mathrm{sqrt}(xdis^2 + ydis^2) \qquad (7\text{-}5)$$

where: dis_h is the total horizontal displacement; $xdis$ and $ydis$ are the displacements in the X direction and Y direction, respectively.

The bottom view of horizontal displacement nephogram slice and vertical displacement nephogram slice of tunnel is shown in Figure 7-22 and Figure 7-23.

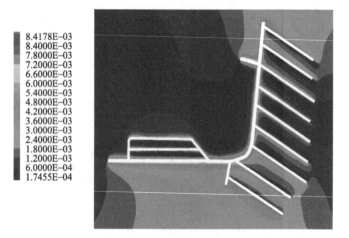

Figure 7-22 Nephogram slice of horizontal displacement

The distribution of horizontal displacement nephogram shows that the maximum horizontal displacement is located at the corner of the slope-connected tunnel and the transportation tunnel outside the vein in north-south direction, and the maximum horizontal displacement at the corner is 8.4178 mm; secondly, the horizontal displacement of the transportation tunnel outside the vein in the east-west direction is higher than the displacement of the transportation tunnel outside the vein and the mining-connected tunnel in the north-south direction, which is basically consistent with the maximum shear stress distribution. The nephogram of roof subsidence shows that the roof subsidence is the largest at the intersection of slope-connected tunnel and mining-connected tunnel, followed by

■ 7 Stability Analysis and Ground Pressure Monitoring ■

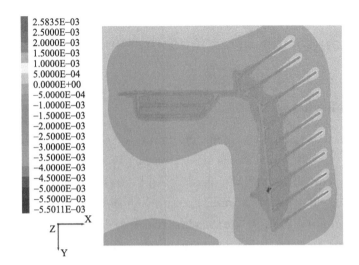

Figure 7-23 Nephogram of roof subsidence

the intersection of mining-connected tunnel and transportation tunnel outside the vein. According to the analysis of the nephogram, the intersection of the slope-connected tunnel and the mining-connected tunnel is in the state of excavation in the east, west, north and south directions, and the vertical stress at the intersection of the slope-connected tunnel and the mining-connected tunnel increases due to the excavation unloading effect, thus increasing the roof subsidence at this place.

The above numerical analysis is based on the assumption of continuous medium. In fact, there are a large number of joints and fissures in the surrounding rock of tunnels. The mechanical properties of the joints and fissures groups largely determine the stability of the surrounding rock of tunnels, and the discontinuity of joints and fissures also affects the stress distribution and deformation of the surrounding rock of tunnels to a large extent. Different combinations of joints and fissures cut the surrounding rock of tunnels, and there are a lot of unstable blocks in the tunnels. The surrounding rock is exposed due to excavation of tunnels, and the stress state changes from three-dimensional to nearly two-dimensional stress state, so the unstable blocks are easy to slip and fall off towards the free surface. When the unstable block on the rock wall of the tunnel slips, it creates a new free surface for the deep block. With the gradual transfer of stress to the deep, the surrounding rock of tunnels is prone to chain failure, and the tunnels will be unstable.

7.4 Monitoring of ground pressure in deep mining

7.4.1 Monitoring method

The deformation, displacement and failure of surrounding rock of tunnels are the result of the internal stress of rock mass. The stress state in rock mass is very complicated, including the dead weight stress of overburden, residual stress of geological structure, water pressure and so on. After the excavation of tunnels, the internal stress in surrounding rock is redistributed, and its distribution and mining technical conditions are closely related to the physical and mechanical properties of rock mass. Drilling stress monitoring is one of the important methods to measure the changes in the internal stress field of the rock strata due to mining, and to study the law of pressure on the stope. It can provide a scientific basis for the stability evaluation of the surrounding rock of tunnels and the optimization of the support design, and has important guiding significance for the safety production of mines.

(1) Monitoring instrument and principle

ZLGH drilling dynamometer and GSJ-2A vibrating wire data technology storage are used (Figure 7-24). ZLGH drilling dynamometer is a kind of vibrating string sensor with special structure. When it is installed and used, it can be set at any position within a certain depth in the drilling according to the needs, and the direction of force measurement can be selected at will, so it is convenient to use, and is usually used with GSJ-2A data storage. The main technical indexes are shown in Table 7-6.

Figure 7-24 ZLGH drilling dynamometer and data acquisition instrument

Main technical indexes of ZLGH drilling dynamometer   Table 7-6

Range	Accuracy	Repeatability	Resolution	Outer diameter
20 MPa, 40 MPa, 60 MPa	0.5% FS, 1.0% FS	0.2% FS, 0.4% FS	0.01% FS	40 mm, 60 mm

(2) Arrangement of measuring points and installation of drilling dynamometer

Drilling rig 7655 is used in the construction of stress meter drilling, with the diameter of 42 mm, the depth of 3 m and 4 m respectively, and the inclination of 3° upward. The layout construction and daily data acquisition of drilling dynamometer are shown in Figure 7-25.

Figure 7-25 On-site monitoring at −780 m level

The following factors shall be considered when determine the layout position of ground pressure monitoring points:

1) There are clear research objects and practical research significance, and the monitoring data can reflect the law of ground pressure activity caused by mining activities;

2) When determine the position of the drilling dynamometer away from the floor, the air pipe and water pipe shall be avoided, so that the normal production of the mine will not be affected in the process of drilling and installation;

3) Under the premise that conditions permit, in order to facilitate the centralized data collection in daily monitoring work, the measuring points of surrounding rock stress monitoring shall be arranged as close as possible to the measuring points of the section measurement. In the process of on-site construction, it is necessary to use the drill to enlarge the hole in the drilling to ensure the smoothness of the drilling;

4) After the completion of construction, the dynamometer shall be arranged

in the drilling hole in strict accordance with the installation operation rules.

7.4.2 Variation of the ground pressure

(1) Monitoring results of stress at −780 m level

The monitoring data of mining stress at −780 m level is shown in Table 7-7, and the mining stress change curve of tunnels at −780 m level is shown in Figure 7-26. Compared with the initial stress, the stress value at −780 m level increases greatly in the fifth acquisition, and the change of disturbance stress does not tend to be stable due to the increase of distance. Therefore, the surrounding rock of tunnels has obvious stress change under the long-term continuous disturbance of blasting and mining operation.

Monitoring data of mining stress at −780 m level   Table 7-7

Monitoring times	Mining stress value (MPa)	Cumulative variation of mining stress(MPa)	Successive variation of mining stress(MPa)
1	18.09	0	0
2	17.92	−0.17	−0.17
3	17.61	−0.48	−0.31
4	17.76	−0.33	0.15
5	17.82	−0.27	0.06
6	18.87	0.78	1.05
7	19.58	1.49	0.71
8	19.54	1.45	−0.04
9	19.5	1.41	−0.04

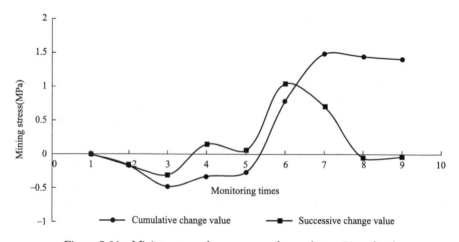

Figure 7-26 Mining stress change curve of tunnel at −780m level

(2) Monitoring results of stress at −795 m level

The location of some measuring points of tunnels at −795 m level is shown in Figure 7-27.

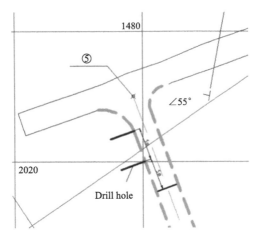

Figure 7-27 Monitoring position at −795 m level

The monitoring data of mining stress at No. 2 measuring point at −795 m level is shown in Table 7-8, and the mining stress change curve of tunnels at No. 2 measuring point at −795 m level is shown in Figure 7-28.

Monitoring data of mining stress at No. 2 measuring point at −795 m level

Table 7-8

Monitoring times	Mining stress value (MPa)	Cumulative variation of mining stress(MPa)	Successive variation of mining stress(MPa)
1	21.66	0	
2	21.58	−0.08	−0.08
3	21.44	−0.22	−0.14
4	21.21	−0.45	−0.23
5	21.34	−0.32	0.13
6	21.51	−0.15	0.17
7	21.69	0.03	0.18
8	21.74	0.08	0.05
9	21.76	0.10	0.02
10	21.75	0.09	−0.01
11	21.74	0.08	−0.01
12	21.74	0.08	0

continued

Monitoring times	Mining stress value (MPa)	Cumulative variation of mining stress(MPa)	Successive variation of mining stress(MPa)
13	21.75	0.09	0.01
14	21.76	0.10	0.01
15	21.78	0.12	0.02
16	21.78	0.12	0
17	21.78	0.12	0
18	21.78	0.12	0
19	21.79	0.13	0.01
20	21.8	0.14	0.01

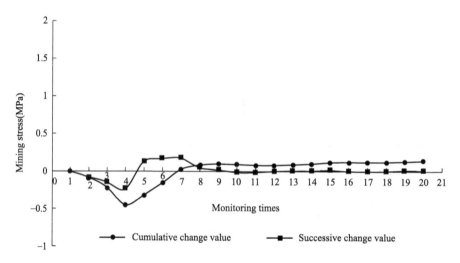

Figure 7-28　Mining stress change curve of tunnels at No. 2 measuring point at −795 m level

The monitoring data of mining stress at No. 3 measuring point at −795 m level is shown in Table 7-9, and the mining stress change curve of tunnels at No. 3 measuring point at −795 m level is shown in Figure 7-29.

Monitoring data of mining stress at No. 3 measuring point at −795 m level

Table 7-9

Monitoring times	Mining stress value (MPa)	Cumulative variation of mining stress(MPa)	Successive variation of mining stress(MPa)
1	21.66	0	
2	22.22	0.56	0.56

7　Stability Analysis and Ground Pressure Monitoring

continued

Monitoring times	Mining stress value (MPa)	Cumulative variation of mining stress(MPa)	Successive variation of mining stress(MPa)
3	22.01	0.35	−0.21
4	21.87	0.21	−0.14
5	21.86	0.20	−0.01
6	21.86	0.20	0
7	21.84	0.18	−0.02
8	21.84	0.18	0
9	21.82	0.16	−0.02
10	21.81	0.15	−0.01
11	21.81	0.15	0
12	21.82	0.16	0.01
13	21.83	0.17	0.01
14	21.83	0.17	0
15	21.81	0.15	−0.02
16	21.83	0.17	0.02
17	21.82	0.16	−0.01
18	21.83	0.17	0.01
19	21.84	0.18	0.01
20	21.84	0.18	0

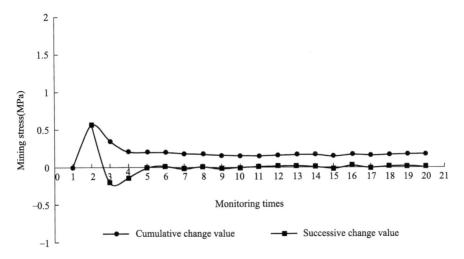

Figure 7-29　Mining stress change curve of tunnels at No. 3 measuring point at −795 m level

The monitoring data of mining stress at No. 6 measuring point at -795 m level is shown in Table 7-10, and the mining stress change curve of tunnels at No. 6 measuring point at -795 m level is shown in Figure 7-30.

Monitoring data of mining stress at No. 6 measuring point at -795 m level

Table 7-10

Monitoring times	Mining stress value (MPa)	Cumulative variation of mining stress(MPa)	Successive variation of mining stress(MPa)
1	21.66	0	
2	22.23	0.57	0.57
3	22.21	0.55	−0.02
4	22.19	0.53	−0.02
5	22.13	0.47	−0.06
6	22.06	0.40	−0.07
7	22.07	0.41	0.01
8	22.03	0.37	−0.04
9	22.05	0.39	0.02
10	22.08	0.42	0.03
11	22.07	0.41	−0.01
12	22.07	0.41	0
13	22.02	0.36	−0.05
14	21.98	0.32	−0.04
15	21.97	0.31	−0.01
16	21.97	0.31	0
17	21.96	0.30	−0.01
18	21.96	0.30	0
19	21.97	0.31	0.01
20	21.94	0.28	−0.03

Mining stress data at -795 m level tends to be stable with the deepening of the tunnel excavation depth and the impact of blasting vibration and rock excavation disturbance on the rock mass stress around the monitoring point gradually decreases with the increase of distance. The two adjacent stress changes gradually tend to 0, and the stress of surrounding rock of tunnels remains in a relatively stable state.

7 Stability Analysis and Ground Pressure Monitoring

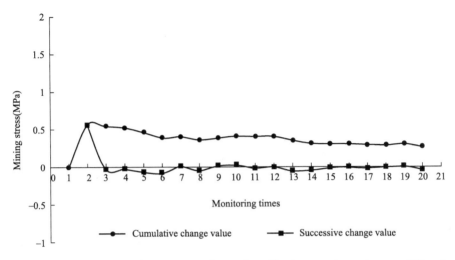

Figure 7-30　Mining stress change curve of tunnels at No. 6 measuring point at −795 m level

(3) Monitoring results of stress at −870 m level

The monitoring of mining stress at −870 m level started in March 2017, and the location of measuring points is shown in the Figure 7-31 and Table 7-11.

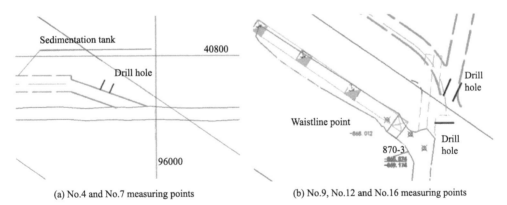

(a) No.4 and No.7 measuring points　　　(b) No.9, No.12 and No.16 measuring points

Figure 7-31　Schematic diagram of measuring points at −870 m level

Layout of monitoring points for mining stress　　　　Table 7-11

Measuring point No.	Level (m)	Position	Drilling hole depth of dynamometer(m)	Distance from the floor(m)
4	−870	Line 1480	2	1.5
7	−870	Line 1480	2	1.5
9	−870	Line 1640	2	1.4

221

Measuring point No.	Level (m)	Position	Drilling hole depth of dynamometer(m)	Distance from the floor(m)
12	−870	Line 1640	2	1.9
16	−870	Line 1640	2	1.4
33	−915	Line 1500	2	1.5

The monitoring data of mining stress at No. 4 and No. 7 measuring points at −870 m level are shown in Table 7-12, and the mining stress change curves of tunnels at No. 4 and No. 7 measuring points at −870 m level are shown in Figure 7-32 and Figure 7-33.

Monitoring data of mining stress at No. 4 and No. 7 measuring points at −870 m level

Table 7-12

Monitoring times	Mining stress of No. 4 measuring point(MPa)	Mining stress of No. 7 measuring point(MPa)
1	0	0
2	2.28	−1.14
3	1.03	−1.1
4	0.3	−1.11
5	−0.05	−1.49
6	−0.23	−1.87
7	−0.18	−1.94
8	0.03	−2.97
9	0.08	−3.23
10	0.1	−3.37
11	−0.32	−3.25
12	−0.32	−3.25
13	−0.32	−3.27
14	−0.32	−3.12
15	−0.32	−3.06
16	−0.33	−3.21
17	−0.32	−3.38
18	−0.32	−3.55
19	−0.26	−3.35

continued

Monitoring times	Mining stress of No. 4 measuring point(MPa)	Mining stress of No. 7 measuring point(MPa)
20	−0.34	−3.42
21	−0.32	−3.37
22	−0.34	−3.41
23	−0.35	−3.51
24	−0.35	−3.51
25	−0.35	−3.5
26	−0.36	−3.48
27	−0.36	−3.45
28	−0.36	−3.42

Note: No. 4 measuring point is vertical stress, and No. 7 measuring point is horizontal stress.

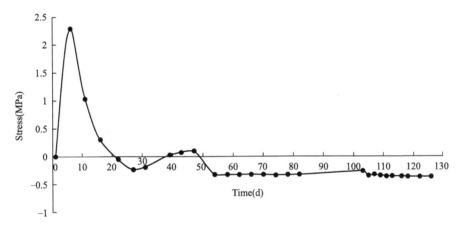

Figure 7-32 Change curve of mining stress at No. 4 measuring point at −870 m level (vertical direction)

No. 4 measuring point is adjacent to No. 7 measuring point, and the stress change of rock mass in the same area is tested. The stress change trend in the vertical direction of No. 4 shows that the stress in the vertical direction of rock mass in this area remains stable after 50 days. The stress in the horizontal direction still has a big change. The negative value of the stress indicates that the expansion of hole has occurred in the drill hole. Because the pre-stress is 5 MPa, the internal stress of the drilling hole is gradually reduced from the initial 5 MPa to 1.58 MPa on June 15th. The surrounding rock mass moves or the stress transfers, resulting in the obvious change of the-stress in the horizontal direc-

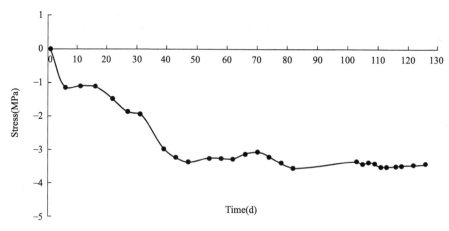

Figure 7-33 Change curve of mining stress at No. 7 measuring point at −870 m level (horizontal direction)

tion, and the rock mass is still in an unstable state.

The monitoring data of mining stress at No. 9, No. 12 and No. 16 measuring points at −870 m level are shown in Table 7-13, and the change curves of mining stress at No. 9, No. 12 and No. 16 measuring points at −870 m level are shown in Figure 7-34 to Figure 7-36.

Monitoring data of mining stress at No. 9, No. 12 and No. 16 measuring points at −870 m level

Table 7-13

Monitoring times	Mining stress of No. 9 measuring point(MPa)	Mining stress of No. 12 measuring point(MPa)	Mining stress of No. 16 measuring point(MPa)
1		0	
2	0	0.8	
3	−0.95	0.5	0
4	−1.04	0.8	−0.41
5	−0.91	0.8	−0.79
6	−1.08	0.8	−0.82
7	−1.14	0.79	−0.83
8	−0.74	0.79	−0.65
9	−1.29	0.79	−0.79
10	−1.41	0.8	−0.87
11	−1.43		−1.21
12	−1.39	Covered with concrete	1.43

continued

Monitoring times	Mining stress of No. 9 measuring point(MPa)	Mining stress of No. 12 measuring point(MPa)	Mining stress of No. 16 measuring point(MPa)
13	−1.42		−2.78
14	−4.76		−2.37
15	−4.76		−2.92
16	−4.77		−2.94
17	−4.77		−2.98
18	−4.77		−3
19	−4.76		−3.01
20	−4.76		−3.03
21	−4.76		−3.04

Note: No. 9 measuring point is vertical stress, No. 12 measuring point is horizontal stress, and No. 16 measuring point is horizontal stress.

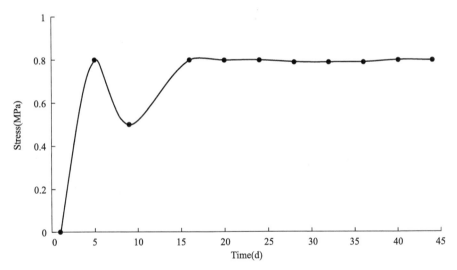

Figure 7-34 Change curve of mining stress at No. 12 measuring point at −870 m level (horizontal direction)

As shown in Figure 7-34, the stress of dynamometer at No. 12 measuring point tends to be stable after 15 days of monitoring.

The stress in the vertical direction of No. 9 changed abruptly after monitoring 30 days and then entered a relatively stable state. The significant change of stress at No. 16 measuring point indicates that the rock mass has undergone deformation or stress transfer, and the surrounding rock mass or rock mass is in an

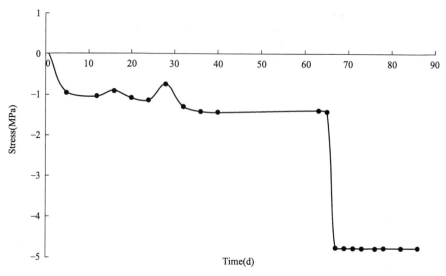

Figure 7-35 Change curve of mining stress at No. 9 measuring point at −870 m level (vertical direction)

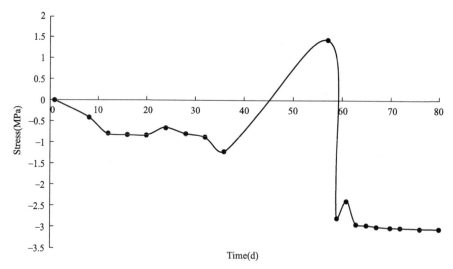

Figure 7-36 Change curve of mining stress at No. 16 measuring point at −870 m level (horizontal direction)

unstable state. In particular, the monitoring data in June has a sharp increase and a steep drop. After the instrument failure problem is eliminated, further monitoring and protection of the place and surrounding rock masses are required.

References

[1] Brown E T. Progress and challenges in some areas of deep mining [J]. Mining Technology, 2012, 121 (4): 177-191.

[2] Cai M, Brown E T. Challenges in the mining and utilization of deep mineral resources [J]. Engineering, 2017, 3 (4): 432-433.

[3] Cai M, Xue D, Ren F. Current status and development strategy of metal mines [J]. Journal of Engineering Science, 2019, 41 (4): 417-426.

[4] Chang J, Xie G X. Mechanical characteristics and stability control of rock roadway surrounding rock in deep mine [J]. Journal of China Coal Society, 2009, 34 (7): 881-886.

[5] Chen L. Reaserch on the energy system of deep granite considering the stored energy and dissipated energy characteristics [D]. Beijing: University of Science and Technology Beijing, 2019.

[6] He M C, Xie H P, Peng S P, et al. Study on rock mechanics in deep mining engineering [J]. Chinese Journal of rock mechanics and engineering, 2005, 24 (16): 2803-2813.

[7] Hu S, Qi C, Zhao S, et al. Discussion on China deep mine classification and critical depth [J]. Coal Science and Technology, 2010, 38 (7): 10-13, 43.

[8] Kang H, Wang J, Lin J. High pretensioned stress and intensive bolting system and its application in deeproadways [J]. Journal of China Coal Society, 2007 (12): 1233-1238.

[9] Li S C, Wang H P, Qian Q H, et al. In-situ monitoring research on zonal disintegration of surrounding rock mass in deep mine roadways [J]. Chinese Journal of Rock Mechanics and Engineering, 2008, 27 (8): 1545-1553.

[10] Li W X, Wen L, Liu X M. Ground movements caused by deep underground mining in Guan-Zhuang iron mine, Luzhong, China [J]. International Journal of Applied Earth Observation and Geoinformation, 2010, 12 (3): 175-182.

[11] Li X, Wang S, Malekian R, et al. Numerical simulation of rock breakage modes under confining pressures in deep mining: an experimental investigation [J]. IEEE Access, 2016, 4: 5710-5720.

[12] Ma N, Zhao X, Zhao Z, et al. Stability analysis and control technology of mine roadway roof in deep mining [J]. Journal of China Coal Society, 2015, 40 (10): 2287-2295.

[13] Meng Q, Han L, Chen Y, et al. Influence of dynamic pressure on deep underground soft rock roadway support and its application [J]. International Journal of Mining Science and Technology, 2016, 26 (5): 903-912.

[14] Ranjith P G, Zhao J, Ju M, et al. Opportunities and challenges in deep mining: a brief review [J]. Engineering, 2017, 3 (4): 546-551.

[15] Wagner H. Deep mining: a rock engineering challenge [J]. Rock Mechanics and Rock Engineering, 2019, 52 (5): 1417-1446.

[16] Wang H, Jiang Y, Xue S, et al. Assessment of excavation damaged zone around roadways under dynamic pressure induced by an active mining process [J]. International Journal of Rock Mechanics and Mining Sciences, 2015, 77: 265-277.

[17] Wang Q, Jiang Z, Jiang B, et al. Research on an automatic roadway formation method in deep mining areas by roof cutting with high-strength bolt-grouting [J]. International Journal of Rock Mechanics and Mining Sciences, 2020, 128: 104264.

[18] Xie H P, Zhou H W, Xue D J, et al. Research and consideration on deep coal mining and critical mining depth [J]. Journal of China Coal Society, 2012, 37 (4): 535-542.

[19] Xie H, Gao F, Ju Y. Research and development of rock mechanics in deep ground engineering [J]. Journal of Rock Mechanics and Engineering, 2015, 34 (11): 2161-2178.

[20] Xie H P, Ju Y, Gao M, et al. Theories and technologies for in-situ fluidized mining of deep underground coal resources [J]. Journal of China Coal Society, 2018, 43 (5): 1210-1219.

[21] Zhou H, Qu C, Hu D, et al. In situ monitoring of tunnel deformation evolutions from auxiliary tunnel in deep mine [J]. Engineering Geology, 2017, 221: 10-15.

[22] Cai M. Prediction and prevention of rockburst in metal mines——A case study of Sanshandao Gold Mine [J]. Journal of Rock Mechanics and Geotechnical Engineering, 2016, 8 (2): 204-211.

[23] Chen Y, Zhao G, Wang S, et al. A case study on the height of a water-flow fracture zone above undersea mining: Sanshandao Gold Mine, China [J]. Environmental Earth Sciences, 2019, 78 (4): 122.

[24] Fan H R, Zhai M G, Xie Y H, et al. Ore-forming fluids associated with granite-hosted gold mineralization at the Sanshandao deposit, Jiaodong gold province, China [J]. Mineralium Deposita, 2003, 38 (6): 739-750.

[25] Gu H, Ma F, Guo J, et al. Assessment of water sources and mixing of groundwater in a coastal mine: the Sanshandao Gold Mine, China [J]. Mine Water and the Environment, 2018, 37 (2): 351-365.

[26] Gu H, Ma F, Guo J, et al. Hydrochemistry, multidimensional statistics, and rock mechanics investigations for Sanshandao Gold Mine, China [J]. Arabian Journal of Geosciences, 2017, 10 (3): 1-17.

[27] Ji D, Xu C, Li T, et al. Hydrogeology investigation and characteristics analysis of seepage field for coastal mine in deep exploitation [J]. Chinese Journal of Engineering Geology, 2016, 24 (4): 674-681.

[28] Li X C, Fan H R, Santosh M, et al. Hydrothermal alteration associated with Mesozoic granite-hosted gold mineralization at the Sanshandao deposit, Jiaodong Gold Province, China [J]. Ore Geology Reviews, 2013, 53: 403-421.

[29] Liu C, Peng B, Qin J. Geological analysis and numerical modeling of mine discharges

for the Sanshandao Gold Mine in Shandong, China: 1. Geological Analysis [J]. Mine Water and the Environment, 2007, 26 (3): 160-165.

[30] Liu Z, Dang W, He X, et al. Cancelling ore pillars in large-scale coastal gold deposit: A case study in Sanshandao Gold Mine, China [J]. Transactions of Nonferrous Metals Society of China, 2013, 23 (10): 3046-3056.

[31] Liu G, Ma F, Zhao H, et al. Study on the fracture distribution law and the influence of discrete fractures on the stability of roadway surrounding rock in the Sanshandao coastal Gold Mine, China [J]. Sustainability, 2019, 11 (10): 2758.

[32] Peng K, Li X, Wang Z. Hydrochemical characteristics of groundwater movement and evolution in the Xinli deposit of the Sanshandao Gold Mine using FCM and PCA methods [J]. Environmental Earth Sciences, 2015, 73 (12): 7873-7888.

[33] Tian L, Zhang Y, Zhang J. Influence of the occurrence status of overburden fault on roadway stability in Sanshandao Gold Mine [J]. Metal Mines, 2018 (1): 178-182.

[34] Wen B J, Fan H R, Hu F F, et al. Fluid evolution and ore genesis of the giant Sanshandao gold deposit, Jiaodong gold province, China: Constrains from geology, fluid inclusions and H-O-S-He-Ar isotopic compositions [J]. Journal of Geochemical Exploration, 2016, 171: 96-112.

[35] Xu W G, Fan H R, Yang K F, et al. Exhaustive gold mineralizing processes of the Sanshandao gold deposit, Jiaodong Peninsula, eastern China: Displayed by hydrothermal alteration modeling [J]. Journal of Asian Earth Sciences, 2016, 129: 152-169.

[36] Zhang L, Groves D I, Yang L Q, et al. Relative roles of formation and preservation on gold endowment along the Sanshandao gold belt in the Jiaodong gold province, China: importance for province-to district-scale gold exploration [J]. Mineralium Deposita, 2020, 55 (2): 325-344.

[37] Abou-Sayed A S, Brechtel C E, Clifton R J. In situ stress determination by hydrofracturing: a fracture mechanics approach [J]. Journal of Geophysical Research: Solid Earth, 1978, 83 (B6): 2851-2862.

[38] Cai M, Qiao L, Yu B, et al. Deep in-situ stress measurement and its distribution law in Jinchuan No. 2 mining area [J]. Journal of Rock Mechanics and Engineering, 1999 (4): 46-50.

[39] Cai M, Qiao L, Li C. Study on in-situ stress field measurement and its distribution law in Xincheng Gold Mine [J]. Nonferrous Metals, 2000 (3): 1-6.

[40] Cai M, Qiao L, Yu B, et al. In-situ stress measurement results and analysis of meishan iron mine [J]. Journal of Rock Mechanics and Engineering, 1997 (3): 34-40.

[41] Cai M F, Liu W D, Li Y. In-situ stress measurement at deep position of Linglong gold mine and distribution law of in-situ stress field in minearea [J]. Chinese Journal of Rock Mechanics and Engineering, 2010, 29 (2): 227-233.

[42] Brown E T, Hoek E. Trends in relationships between measured in-situ stresses and depth [C] //International Journal of Rock Mechanics and Mining Sciences & Geome-

chanics Abstracts. Pergamon, 1978, 15 (4): 211-215.

[43] Haimson B, Fairhurst C. In-situ stress determination at great depth by means of hydraulic fracturing [C] //The 11th US symposium on rock mechanics (USRMS). American Rock Mechanics Association, 1969.

[44] Kang H, Zhang X, Si L, et al. In-situ stress measurements and stress distribution characteristics in underground coal mines in China [J]. Engineering Geology, 2010, 116 (3-4): 333-345.

[45] Li P, Cai M, Guo Q, et al. In situ stress state of the northwest region of the Jiaodong Peninsula, China from overcoring stress measurements in three gold mines [J]. Rock Mechanics and Rock Engineering, 2019, 52 (11): 4497-4507.

[46] Li P, Ren F, Cai M, et al. Present-day stress state and fault stability analysis in the capital area of China constrained by in situ stress measurements and focal mechanism solutions [J]. Journal of Asian Earth Sciences, 2019, 185: 104007.

[47] Li P, Cai M, Miao S, et al. New Insights Into The Current Stress Field Around the Yishu Fault Zone, Eastern China [J]. Rock Mechanics and Rock Engineering, 2019, 52 (10): 4133-4145.

[48] Li P, Cai M, Guo Q, et al. Characteristics and implications of stress state in a gold mine in Ludong area, China [J]. International Journal of Minerals, Metallurgy, and Materials, 2018, 25 (12): 1363-1372.

[49] Li P, Cai M. Distribution law of in situ stress field and regional stress field assessments in the Jiaodong Peninsula, China [J]. Journal of Asian Earth Sciences, 2018, 166: 66-79.

[50] Li Y. In-situ stress measurement and stability analysis based on the unified strength theory in large scale underground cavernszone [D]. Beijing: University of Science and Technology Beijing, 2008.

[51] Lim S S, Martin C D, Åkesson U. In-situ stress and microcracking in granite cores with depth [J]. Engineering geology, 2012, 147: 1-13.

[52] Pine R J, Ledingham P, Merrifield C M. In-situ stress measurement in the Carnmenellis granite—II. Hydrofracture tests at Rosemanowes quarry to depths of 2000 m [C] //International Journal of Rock Mechanics and Mining Sciences & Geomechanics Abstracts. Pergamon, 1983, 20 (2): 63-72.

[53] Peng S, Meng Z. Theory and Practice of Mine Engineering Geology [M]. Beijing: Geological Publishing House, 2002.

[54] Shao A. Coal Mine Groundwater [M]. Beijing: Geological Publishing House, 2005.

[55] Stephansson O, Zang A. ISRM suggested methods for rock stress estimation—part 5: establishing a model for the in situ stress at a given site [J]. Rock Mechanics and Rock Engineering, 2012, 45 (6): 955-969.

[56] Wang S, Cai M, Miao S, et al. Results and their analysis of in situ stress measurement in Sanshandao Gold Mine [J]. China Mining Industry, 2003 (10): 46-48.

[57] Yan Z. Study on Digital Mining Technique and Mining Index Optimization of Sublevel Caving in Gongchangling lron Mine [D]. Beijing: University of Science and Technology Beijing, 2019.

[58] Zhang Z, Cai M. The effect of underground water on ground stress measurement [J]. Metal Mines, 2001 (11): 42-44.

[59] Zhao X G, Wang J, Cai M, et al. In-situ stress measurements and regional stress field assessment of the Beishan area, China [J]. Engineering geology, 2013, 163: 26-40.

[60] Zoback M D, Haimson B C. Status of the hydraulic fracturing method for in-situ stress measurements [C] //The 23rd US Symposium on Rock Mechanics (USRMS). American Rock Mechanics Association, 1982.

[61] Basahel H, Mitri H. Application of rock mass classification systems to rock slope stability assessment: A case study [J]. Journal of rock mechanics and geotechnical engineering, 2017, 9 (6): 993-1009.

[62] Cai B, Yu Y, Wu X. Relationship between engineering rock mass classification standard, Q classification and RMR classification and estimation of deformation parameters [J]. Journal of Rock Mechanics and Engineering, 2001 (S1): 1677-1679.

[63] Cai M, Kaiser P. Visualization of rock mass classificationsystems [J]. Geotechnical & Geological Engineering, 2006, 24 (4): 1089-1102.

[64] Cao J, Ma F, Guo J, et al. Assessment of mining-related seabed subsidence using GIS spatial regression methods: a case study of the Sanshandao Gold Mine (Laizhou, Shandong Province, China) [J]. Environmental Earth Sciences, 2019, 78 (1): 26.

[65] Chao J, Shufang X, Ning L. Application of extenics theory to evaluation of tunnel rock quality [J]. Chinese Journal of Rock Mechanics and Engineering, 2003, 22 (5): 751-756.

[66] Cao W G, Zhai Y C, Wang J Y. Combination evaluation method for classification of surrounding rock quality of tunnel based on drifting degree [J]. Chinese Journal of Geotechnical Engineering, 2012, 34 (6): 978-983.

[67] Chen C, Wang G. Discussion on the interrelation of various rock mass quality classification systems at home and abroad [J]. Journal of Rock Mechanics and Engineering, 2002 (12): 1894-1900.

[68] Deere D. The rock quality designation (RQD) index in practice [M] //Rock classification systems for engineering purposes. West Conshohocken: ASTM International, 1988.

[69] Han X, Chen J, Wang Q, et al. A 3D fracture network model for the undisturbed rock mass at the Songta dam site based on small samples [J]. Rock Mechanics and Rock Engineering, 2016, 49 (2): 611-619.

[70] Haneberg W C. Using close range terrestrial digital photogrammetry for 3-D rock slope modeling and discontinuity mapping in the United States [J]. Bulletin of Engineering Geology and the Environment, 2008, 67 (4): 457-469.

[71] Hagan T O. A case for terrestrial photogrammetry in deep-mine rock structure studies

[C] //International Journal of Rock Mechanics and Mining Sciences & Geomechanics Abstracts. Pergamon, 1980, 17 (4): 191-198.

[72] Huang N, Jiang Y, Liu R, et al. A novel three-dimensional discrete fracture network model for investigating the role of aperture heterogeneity on fluid flow through fractured rock masses [J]. International Journal of Rock Mechanics and Mining Sciences, 2019, 116: 25-37.

[73] Liang G L, Xu W Y, Tan X L. Application of extension theory based on entropy weight to rock quality evaluation [J]. Rock and Soil Mechanics, 2010, 31 (2): 535-540.

[74] Liu Z, Dang W. Rock quality classification and stability evaluation of undersea deposit based on M-IRMR [J]. Tunnelling and Underground Space Technology, 2014, 40: 95-101.

[75] Palmstrom A. Measurements of and correlations between block size and rock quality designation (RQD) [J]. Tunnelling and Underground Space Technology, 2005, 20 (4): 362-377.

[76] Qi J. Study on failure mechanism and control technology of surrounding rock of deeptunnel in xishimen iron mine [D]. Beijing: University of Science and Technology Beijing, 2018.

[77] Singh B, Goel R K. Rock mass classification: a practical approach in civil engineering [M]. Amsterdam: Elsevier, 1999.

[78] Şen Z. Rock quality designation-fracture intensity index method for geomechanical classification [J]. Arabian Journal of Geosciences, 2014, 7 (7): 2915-2922.

[79] Wang M, Zhang N, Li J, et al. Computational method of large deformation and its application in deep mining tunnel [J]. Tunnelling and Underground Space Technology, 2015, 50: 47-53.

[80] Wang S, Zhou H, WU C, et al. Research on rock mechanics parameters by using comprehensive evaluation method of rock quality grade-oriented [J]. Rock and Soil Mechanics, 2007, 28 (S1): 202-206.

[81] Wang S, Ni P, Guo M. Spatial characterization of joint planes and stability analysis of tunnel blocks [J]. Tunnelling and Underground Space Technology, 2013, 38: 357-367.

[82] Wang J, Guo J. Research on Rock Mass Quality Classification Based on An Improved Rough Set-Cloud Model [J]. IEEE Access, 2019, 7: 123710-123724.

[83] Wu L Z, Li S H, Zhang M, et al. A new method for classifying rock mass quality based on MCS and TOPSIS [J]. Environmental Earth Sciences, 2019, 78 (6): 1-11.

[84] Yang C, Luo Z, Hu G, et al. Application of a microseismic monitoring system in deep mining [J]. Journal of University of Science and Technology Beijing, Mineral, Metallurgy, Material, 2007, 14 (1): 6-8.

[85] Yong S, Changhong Y, Qinghua M. Comprehensive application of rock quality evaluation system [J]. Chinese Journal of Underground Space and Engineering, 2016, 12 (4): 1129-1134.

[86] Young R P, Coffey J R, Hill J J. The application of spectral analysis to rock quality evaluation for mapping purposes [J]. Bulletin of the International Association of Engineering Geology-Bulletin de l'Association Internationale de Géologie de l'Ingénieur, 1979, 19 (1): 268-274.

[87] Zhang L. Determination and applications of rock quality designation (RQD) [J]. Journal of Rock Mechanics and Geotechnical Engineering, 2016, 8 (3): 389-397.

[88] Zheng J, Wang X, Lü Q, et al. A Contribution to Relationship Between Volumetric Joint Count (J v) and Rock Quality Designation (RQD) in Three-Dimensional (3-D) Space [J]. Rock Mechanics and Rock Engineering, 2020, 53 (3): 1485-1494.

[89] Zheng J, Yang X, Lü Q, et al. A new perspective for the directivity of rock quality designation (RQD) and an anisotropy index of jointing degree for rockmasses [J]. Engineering geology, 2018, 240: 81-94.

[90] Banks D, Younger P L, Arnesen R T, et al. Mine-water chemistry: the good, the bad and the ugly [J]. Environmental Geology, 1997, 32 (3): 157-174.

[91] Feng X T, Li S J, Chen S L. Effect of water chemical corrosion on strength and cracking characteristics of rocks-a review [J]. Key Engineering Materials, 2004, 261: 1355-1360.

[92] Feng X T, Ding W X. Coupled chemical stress processes in rock fracturing [J]. Materials Research Innovations, 2011, 15 (sup1): s547-s550.

[93] Fernando W A M, Ilankoon I, Syed T H, et al. Challenges and opportunities in the removal of sulphate ions in contaminated mine water: A review [J]. Minerals Engineering, 2018, 117: 74-90.

[94] Guo Q, Yan B, Cai M, et al. Permeability Coefficient of Rock Mass in Underwater Mining [J]. Geotechnical and Geological Engineering, 2020, 38 (2): 2245-2254.

[95] Kalin M. Passive mine water treatment: the correct approach? [J]. Ecological Engineering, 2004, 22 (4-5): 299-304.

[96] Karfakis M G, Akram M. Effects of chemical solutions on rock fracturing [C] //International journal of rock mechanics and mining sciences & geomechanics abstracts. Pergamon, 1993, 30 (7): 1253-1259.

[97] Lasaga A C. Chemical kinetics of water-rock interactions [J]. Journal of geophysical research: solid earth, 1984, 89 (B6): 4009-4025.

[98] Liu Y C, Chen C S. A new approach for application of rock mass classification on rock slope stability assessment [J]. Engineering geology, 2007, 89 (1-2): 129-143.

[99] Liu Q. Study on the Flocculation Parameters of Fine Sediments and the Environmental Effects in the Changjiang Estuary [D]. Shanghai: East China Normal University, 2007.

[100] Liu F. Research on groundwater circulation and hydrochemical transport in the Northern part of Ordos Cretaceous Basin based on isotope technology [D]. Changchun: Jilin University, 2008.

[101] Lin Y, Zhou K, Gao F, et al. Damage evolution behavior and constitutive model of

sandstone subjected to chemical corrosion [J]. Bulletin of Engineering Geology and the Environment, 2019, 78 (8): 5991-6002.

[102] Lin Y, Zhou K, Li J, et al. Weakening laws of mechanical properties of sandstone under the effect of chemical corrosion [J]. Rock Mechanics and Rock Engineering, 2020, 53 (4): 1857-1877.

[103] McDermott C, Bond A, Harris A F, et al. Application of hybrid numerical and analytical solutions for the simulation of coupled thermal, hydraulic, mechanical and chemical processes during fluid flow through a fractured rock [J]. Environmental Earth Sciences, 2015, 74 (12): 7837-7854.

[104] Miao S, Wang H, Guo X, et al. The effects of mineralization on the lognormal distribution and exponential fluctuations of a hydrothermal gold deposit in Jiaodong Peninsula, China [J]. Arabian Journal of Geosciences, 2018, 11 (18): 1-11.

[105] Miao S, Wang H, Cai M, et al. Damage constitutive model and variables of cracked rock in a hydro-chemical environment [J]. Arabian Journal of Geosciences, 2018, 11 (2): 1-14.

[106] Miao S, Cai M, Guo Q, et al. Damage effects and mechanisms in granite treated with acidic chemical solutions [J]. International Journal of Rock Mechanics and Mining Sciences, 2016, 88: 77-86.

[107] Miao S J, Cai M F, Ji D, et al. Aging features and mechanism of granite's damage under the action of acidic chemical solutions [J]. Journal of China Coal Society, 2016, 41 (5): 1137-1144.

[108] Miao S J, Cai M F, Ji D. Damage effect of granite's mechanical properties and parameters under the action of acidicsolutions [J]. Journal of China Coal Society, 2016, 41 (4): 829-835.

[109] Palmstrom A, Broch E. Use and misuse of rock mass classification systems with particular reference to the Q-system [J]. Tunnelling and underground space technology, 2006, 21 (6): 575-593.

[110] Singh B, Goel R K. Engineering rock mass classification: tunneling, foundations, andl andslides [M]. Waltham, MA: Butterworth-Heinemann, 2011.

[111] Singh G. Impact of coal mining on mine water quality [J]. International journal of mine water, 1988, 7 (3): 49-59.

[112] Yin S, Zhang J, Liu D. A study of mine water inrushes by measurements of in situ stress and rock failures [J]. Natural Hazards, 2015, 79 (3): 1961-1979.

[113] Zhao Y, Tang L, Liu Q, et al. The Micro Damage Model of the Cracked Rock Considering Seepage Pressure [J]. Geotechnical and Geological Engineering, 2019, 37 (2): 965-974.

[114] Cai M, Ji D, Guo Q. Study of rockburst prediction based on in-situ stress measurement and theory of energy accumulation caused by Mining disturbance [J]. Chinese Journal of Rock Mechanics and Engineering, 2013, 32 (10): 1973-1980.

[115] Gao F, Stead D, Kang H. Numerical simulation of squeezing failure in a coal mine roadway due to mining-induced stresses [J]. Rock Mechanics and Rock Engineering, 2015, 48 (4): 1635-1645.

[116] Grice T. Underground mining with backfill [J]. 2nd Annual Summit-Mine Tailings Disposal Systems, Brisbane, Nov, 1998: 24-25.

[117] Guo G, Zhu X, Zha J, et al. Subsidence prediction method based on equivalent mining height theory for solid backfilling mining [J]. Transactions of Nonferrous Metals Society of China, 2014, 24 (10): 3302-3308.

[118] Guo W, Xu F. Numerical simulation of overburden and surface movements for Wongawilli strip pillar mining [J]. International Journal of Mining Science and Technology, 2016, 26 (1): 71-76.

[119] Jiang J Q, Wang P, Jiang L S, et al. Numerical simulation on mining effect influenced by a normal fault and its induced effect on rock burst [J]. Geomechanics & engineering, 2018, 14 (4): 337-344.

[120] Ju F, Huang P, Guo S, et al. A roof model and its application in solid backfilling mining [J]. International Journal of Mining Science and Technology, 2017, 27 (1): 139-143.

[121] Lu J. Alternate Room-and-pillar Mining with Ascending Backfill Method and its Industrial Application [D]. Changsha: Central South University, 2012.

[122] Li X, Wang S, Malekian R, et al. Numerical simulation of rock breakage modes under confining pressures in deep mining: an experimental investigation [J]. IEEE Access, 2016, 4: 5710-5720.

[123] Ma D, Zhang J, Duan H, et al. Reutilization of gangue wastes in underground backfilling mining: overburden aquifer protection [J]. Chemosphere, 2021, 264: 128400.

[124] Shnorhokian S, Mitri H S, Thibodeau D. Numerical simulation of pre-mining stress field in a heterogeneous rockmass [J]. International Journal of Rock Mechanics and Mining Sciences, 2014 (66): 13-18.

[125] Wang A, Ma L, Wang Z, et al. Soil and water conservation in mining area based on ground surface subsidence control: development of a high-water swelling material and its application in backfilling mining [J]. Environmental Earth Sciences, 2016, 75 (9): 779.

[126] Wei X, Feng Z, Zhao Y. Numerical simulation of thermo-hydro-mechanical coupling effect in mining fault-mode hot dry rock geothermal energy [J]. Renewable Energy, 2019, 139: 120-135.

[127] Yan-Li H, Ji-Xiong Z, Qiang Z. Strata movement control due to bulk factor of backfilling body in fully mechanized backfilling mining face [J]. Journal of Mining and Safety Engineering, 2012, 29 (2): 162.

[128] Yanli H, Jixiong Z, Baifu A, et al. Overlying strata movement law in fully mechanized coal mining and backfilling longwall face by similar physical simulation [J]. Jour-

nal of Mining Science, 2011, 47 (5): 618-627.

[129] Yang Z, Zhai S, Gao Q, et al. Stability analysis of large-scale stope using stage subsequent filling mining method in Sijiaying iron mine [J]. Journal of Rock Mechanics and Geotechnical Engineering, 2015, 7 (1): 87-94.

[130] Zhang J, Zhou N, Huang Y, et al. Impact law of the bulk ratio of backfilling body to overlying strata movement in fully mechanized backfilling mining [J]. Journal of Mining Science, 2011, 47 (1): 73-84.

[131] Zhang J, Zhang Q, Spearing A J S S, et al. Green coal mining technique integrating mining-dressing-gas draining-backfilling-mining [J]. International Journal of Mining Science and Technology, 2017, 27 (1): 17-27.

[132] Zhou S. Research on Stability of Filling and Intelligent Matching between Filling and Rock Mass in Deep Mining [D]. Changsha: Central South University, 2012.

[133] He S, Wang H, Guo Y, et al. Analysis of Characteristics of Seepage Field in Sanshandao Island Gold Mine under the Action of Water Chemical Damage [J]. Metal Mine, 2016 (6): 167-172.

[134] He H, Dou L, Fan J, et al. Deep-hole directional fracturing of thick hard roof for rockburst prevention [J]. Tunnelling and Underground Space Technology, 2012, 32: 34-43.

[135] He J, Dou L, Cao A, et al. Rock burst induced by roof breakage and its prevention [J]. Journal of Central South University, 2012, 19 (4): 1086-1091.

[136] Li W, Qi W, Wu X. Research on surrounding rock stress evolution in deep mining and its numerical simulation based on borehole stress monitoring [J]. Gold, 2019, 40 (5): 27-32.

[137] Li W, Wang H, Wu S. Numerical Simulation and Analysis of Depressure Release Hole in Deep Mining of Sanshandao Gold Mine [J]. Modern Mining, 2019, 35 (9): 98-102.

[138] Lu Y, Su J, Li X. Prevention Measures for Roof Caving in Orebody of Deep Broken Zone [J]. Journal of Mining & Safety Engineering, 2009, 26 (1): 60-64, 69.

[139] Liu H, Zhu M, Wu Q, et al. Determination and monitoring of dangerous body of cracked surrounding rock in deep roadway [J]. China Mining Magazine, 2020, 29 (1): 137-140.

[140] Liu G, Ma F, Zhao H, et al. Study on the fracture distribution law and the influence of discrete fractures on the stability of roadway surrounding rock in the Sanshandao coastal Gold Mine, China [J]. Sustainability, 2019, 11 (10): 2758.

[141] Liu B. Research on the Safety Thickness of Cave Roof under Tunnel based on Equivalent Rock Mechanics Parameters [D]. Beijing: University of Science and Technology Beijing, 2016.

[142] Lou Y, Wu Z, Sun W, et al. Study on failure models and fractal characteristics of shale under seepage-stress coupling [J]. Energy Science & Engineering, 2020, 8 (5): 1634-1649.

[143] Song W, Liang Z. Investigation on failure characteristics and water inrush risk of inclined floor mining above confined aquifer [J]. Rock and Soil Mechanics, 2020, 41 (2): 624-634.

[144] Tian L, Zhang Y, Zhang J. Numerical Simulation and Application of Boreholes Destressing Technology in High In-situ Sress Zone During Deep Mining [J]. Metal Mine, 2017 (4): 31-35

[145] Wang P, Jiang L, Zheng P, et al. Inducing mode analysis of rock burst in fault-affected zone with a hard-thick stratum occurrence [J]. Environmental Earth Sciences, 2019, 78 (15): 1-13.

[146] Xu C, Fu Q, Cui X, et al. Apparent-depth effects of the dynamic failure of thick hard rock strata on the underlying coal mass during underground mining [J]. Rock mechanics and rock engineering, 2019, 52 (5): 1565-1576.

[147] Xia B, Zhang X, Yu B, et al. Weakening effects of hydraulic fracture in hard roof under the influence of stressarch [J]. International Journal of Mining Science and Technology, 2018, 28 (6): 951-958.

[148] Yang T H, Jia P, Shi W H, et al. Seepage-stress coupled analysis on anisotropic characteristics of the fractured rock mass around roadway [J]. Tunnelling and underground space technology, 2014, 43: 11-19.

[149] Zhao J, Yin L, Guo W. Stress-seepage coupling of cataclastic rock masses based on digital image technologies [J]. Rock Mechanics and Rock Engineering, 2018, 51 (8): 2355-2372.

[150] Zhang T, Zhao Y, Gan Q, et al. Investigations into mining-induced stress-fracture-seepage field coupling in a complex hydrogeology environment: a case study in the bulianta colliery [J]. Mine water and the environment, 2019, 38 (3): 632-642.